Paleoclimatology

Paleoclimatology

Edited by:

Vasant Deshmukh

Delve
PUBLISHING

www.delvepublishing.com

Paleoclimatology

Vasant Deshmukh

Delve Publishing
224 Shoreacres Road
Burlington, ON L7L 2H2
Canada
www.delvepublishing.com
Email: orders@arclereducation.com

ISBN: 978-1-77469-138-0 (Hardcover)

ABOUT THE EDITOR

Dr. Vasant A Deshmukh did his under graduation with first class in Agriculture 2003, Masters in Agronomy with First class in 2005 and Doctoral studies with distinction in Agronomy in 2013 at V.N.M.K.V. Parbhani (MS). He served as Assistant professor at various colleges for 11 years for teaching Agronomy and Agricultural meteorology to under graduate students. Currently working as Programme Coordinator at KVK MGM Gandheli Aurangabad. He his 14 research paper with many technical papers to his credit. Actively engaged in trainings of farmers, rural youth and farm women and also act as a resource person in farm extension activities organized by different agencies. Published extension booklets, technical folders, articles, and posters for transfer of technology. Delivered TV talks and radio talks on technology transfer for farming community.

TABLE OF CONTENTS

List of Figures ..... *xi*

List of Tables ..... *xvii*

List of Abbreviations ..... *xix*

Preface ..... *xxi*

Chapter 1 Fundamentals of Paleoclimatology and Climate Change 1

1.1. Introduction 2

1.2. Paleoclimatology: Historical Climate 3

1.3. Paleoclimatic Methods 5

1.4. The Pre-Quaternary Climates 10

1.5. Glacial Interglacial Dynamics and Variability 15

References 32

Chapter 2 History of Paleoclimatology 45

2.1. Introduction 46

2.2. Beginning of Reason 47

2.3. The Scientific Strategy 47

2.4. Concept of Climatic Cycles 49

2.5. Commercialization of Paleoclimatology 53

2.6. Actualism and Uniformitarianism 54

2.7. The Theory of Glacial 55

2.8. From Accumulated Ice to Glaciation 56

2.9. Pre-Quaternary Eras of ICE 57

2.10. Eventually: Climatic Variations 59

2.11. Cosmic Variables 60

2.12. Elements in the Galactic Arms 61

2.13. Solar Radiation 61

2.14. Orbital Dissimilarities 62

2.15. Events of Asteroid Collision 63
2.16. Endogenic Global Heating 65
2.17. Plate Tectonic and Relocation of Oceans and Continents 66
2.18. Orographic Reaction to Eustasy and Plate Tectonics 68
2.19. Uplift of Mountain and Global Cooling 69
2.20. Oceanic Circulation 70
2.21. Sea Ice and the Glacier Variation 71
2.22. Variations in Atmospheric Composition 72
2.23. The Moon, Lunar-Solar Tides, and Associated Solar, Terrestrial, and Planetary Cycles 74
2.24. Volcanic Eruptions and the Desert Dust 77
2.25. Paleosols and Loess 78
2.26. Eolianite and Eolization 79
2.27. Levels of the Lake and Fluvial Signs 80
References 83

Chapter 3 Basics of Paleobotany 95
3.1. Introduction 96
3.2. Nearest-Living-Relative Technique 97
3.3. Plant Structure Proxies 99
3.4. Leaves 100
3.5. Wood 103
3.6. Chemical Signatures 105
References 107

Chapter 4 Paleocean Modeling and Paleoceanography 111
4.1. Introduction 112
4.2. Paleocean Modeling Challenges 113
4.3. Setting Up Deep Paleocean Model Investigations 113
4.4. Prognosis of Paleocean Modeling 116
4.5. Meridional Overturning Circulation (MOC) and Deep-Water Formation 118
4.6. Heat Transport and Ocean Gateways 119
4.7. Paleoceanography 121
4.8. Paleoceanographic Techniques 122

4.9. Paleoceanophic Processes 125
4.10. Paleocene-Eocene Thermal Maximum (PETM) 128
References 137

Chapter 5 Pliocene Climates 147
5.1. Introduction 148
5.2. Climate Dynamics and Climate Modeling 150
5.3. Causes of Pliocene Climate Change Climatic Warmth 156
References 159

Chapter 6 Paleotemperatures and Proxy Reconstructions 165
6.1. Introduction 166
6.2. Methods of Paleotemperature Reconstruction 167
6.3. Other Methods 176
References 177

Chapter 7 Holocene Climate 181
7.1. Introduction 182
7.2. Data Sources 183
7.3. Climate Change During the Latest Deglacial Period 184
7.4. The Climatic Optimum of the Earlier Holocene 185
7.5. Climate Variability in the Late Holocene 187
7.6. Variation of Holocene Treeline 190
References 197

Chapter 8 Human Evolution and Climate Change 203
8.1. Introduction 204
8.2. The Family Tree of Hominins 205
8.3. Hominin Attributes' Evolution 207
8.4. Worldwide Climate Change 209
8.5. Climate and Evolution 210
8.6. What Does Paleoclimatology Teach Us? 212
8.7. Conclusions and Summary 213
References 214

Index 219

LIST OF FIGURES

Figure 1.1. Illustration of weather versus climate.

Figure 1.2. Worldwide average temperatures for the past 540 million years, utilizing graphs from numerous studies compressed together. This graph shows a great deal of time, however at a very low resolution.

Figure 1.3. Earth's various systems.

Figure 1.4. Illustration of climate models.

Figure 1.5. Diagram of the paleoclimate modeling.

Figure 1.6. (Highest) Atmospheric carbon dioxide and continental glaciation of 400 Ma to current. (Central) Universal compiling of profound sea benthic foraminifera 180 isotope chronicles from 40 profound sea boring program and ocean boring program locations (Zachos et al., 2001) modernized through great resolution chronicles for the Eocene by Miocene break (Billups et al., 2002; Zachos and Bohaty, 2003; Lear et al., 2004). (Lowest) Comprehensive record of carbon dioxide for previous 65 Myr. Separate chronicles of carbon dioxide and related issues are color-coded through proxy technique; when probable, chronicles are founded on duplicate examples (Royer, 2006).

Figure 1.7. The PETM as chronicled in benthic (lowest dwelling) foraminifer (Nuttallides truempyi) isotopic chronicles from locations in the Pacific, South Atlantic, and Antarctic (Zachos et al., 2003).

Figure 1.8. Record of glacial interglacial cycles.

Figure 1.9. Differences of deuterium (δD; black), an indigenous temperature proxy, and the atmospheric absorptions of gases of the greenhouse carbon dioxide (red), methane (blue), and nitrous oxide (N_2O; green) taken from air stuck in cores of ice from Antarctica and latest measurements of atmospheric (Petit et al., 1999; Indermühle et al., 2000; EPICA Members of the Community, 2004; Spahni et al., 2005; Siegenthaler et al., 2005a, b).

Figure 1.10. The radiative compelling and concentrations by (a) carbon dioxide; (b) methane; (c) nitrous oxide (N_2O); and (d) the alteration rate in their joined radiative forcing above the previous 20 kyr restored from Greenland and Antarctic ice and records (symbols) and measurements of direct atmosphere (magenta and red lines).

Figure 1.11. Diagram of the Earth's orbital variations (Milankovitch cycles) that led to the cycles of the ice age. 'T' means variations in the tilt (or obliquity) of the axis of Earth, 'E' means variations in the eccentricity of the orbit (because of differences in the slight axis of the ellipse), and 'P' means precession, that is, alterations in the path of the axis tilt at a specified point of the orbit (Schellnhuber and Rahmstorf, 2006).

Figure 1.12. The LGM climate (nearly 21 ka) is comparative to the pre-industrialized (1750) climate.

Figure 1.13. (A) Ice comparable of eustatic sea-level past above the cycle of the previous glacial-interglacial as per the Waelbroeck et al. (2002) examination. (B) Fitting of the ICE-5G(VM2) model forecast (red line) to the extensive inclined coral founded record of RSL history from the Barbados island in the Caribbean Sea (Fairbanks, 1989; Fairbanks and Peltier, 2006) above the age extent from 32 ka to current.

Figure 2.1. Cycles of the flow of the Nile during 10 years (on top) and the complete year (on bottom).

Figure 2.2. Sunspot observation data for the last four years.

Figure 2.3. Importance of paleoclimatology.

Figure 2.4. Phases of the glacial control theory.

Figure 2.5. Chronology of glaciation.

Figure 2.6. Causal loop representation of climate variation.

Figure 2.7. Consequences of cosmic rays on climatic changes.

Figure 2.8. Consequences of solar radiation (energy emissions) on climatic changes.

Figure 2.9. Cycles of Milankovitch.

Figure 2.10. Outline of events of asteroid collision.

Figure 2.11. Endogenic procedures and their cataloging.

Figure 2.12. Schematic representation of the plate tectonics.

Figure 2.13. Mountain uplift of the European alps.

Figure 2.14. Map of the cycles of oceanic circulation.

Figure 2.15. A yearly discrepancy of the arctic sea-ice amount over the years.

Figure 2.16. Fluctuations in the concentration of the greenhouse gases from zero to 2005 A.D.

Figure 2.17. Processes of the lunar-solar tides.

Figure 2.18. Effects of volcanic eruptions on climate fluctuation.

Figure 2.19. Picture presentation loess and paleosols.

Figure 2.20. Profile of longitudinal deposition displaying a progressive downstream reduction in the slope gradient. Loss and gain of lodging space for gathering of fluvial accumulations during an increase in the base level and corrosion and creation of notched valleys during the fall in base levels.

Figure 3.1. With the Chinle group as an example, some basic ideas of paleobotanical biostratigraphy and biochronology.

Figure 3.2. From southern Pennsylvania a hypothetical list of modern relatives based on flora.

Figure 3.3. Percentage of whole margined leaves in a flora versus mean annual temperature for 177 sampled floras global (Wolfe, 1993; Gregory-Wodzicki, 2000).

Figure 3.4. For three sites in China, the ^{13}C isotopic composition of current Ginkgo leaves was compared to yearly precipitation. (Sun et al., 2003).

Figure 4.1. Boreal winter paleoceanography that is simulated using the COSMOS model.

Figure 4.2. Visionary plan of soft computing models for the prediction of sea surface temperature.

Figure 4.3. Barotropic stream function generated by the ocean model (see right figure), and the general and theoretical wind-driven circulation model using ocean circulations (refers to left figure).

Figure 4.4. Cretaceous paleotemperature predictions presented from DSDP and ODP cores dependent upon deep-dwelling fauna are inadequate and show a broad range of values, but fix to a too warmer deep ocean.

Figure 4.5. Climate change and oceanography.

Figure 4.6. Paleotidal scheme of the Burdigalian Peri-Alpine Seaway.

Figure 4.7. In marine carbonate sediments, the assemblage of $d^{18}O$ values as the cretaceous displaying the PETM irregularity in the context of Cenozoic variability (Zachos et al., 2001).

Figure 4.8. Thorough oxygen isotopic series and benthic foraminiferal carbon across the PETM occurrence from ODP stations 865 (Allison Guyot, Equatorial Pacific), 690 (Maud Rise, South Atlantic), 527 and 525 (Walvis Ridge, South Atlantic).

Figure 5.1. Climate model estimation: As a yearly average, the difference between today's surface temperature (C) and simulated Pliocene (3 million years ago) (Haywood et al., 2000a; Haywood, 2001).

Figure 5.2. As a function of latitude, average yearly surface temperatures (C) plotted for present-day (solid line) and model-predicted Pliocene (dashed line). This plot shows a surface temperature region with larger equability and decreased seasonality at the time of the Pliocene relative to modern (Haywood, 2001).

Figure 5.3. As a function of latitude, yearly average total precipitation rate for Pliocene in mm day^{-1} and the present-day (Haywood et al., 2000a; Haywood, 2001).

Figure 5.4. Between the Pliocene and current-day for (a) DJF; and (b) JJA, model-predicted alteration in average total precipitation rate (mm day^{-1}). It is observed that the areas where the variation in precipitation is less than 0.5 mm day^{-1} are not shaded and the contour interval is not equally set apart.

Figure 5.5. Between the Pliocene and current-day for (a) DJF; and (b) JJA, model-predicted alteration in surface pressure (hPa): Vector surface average winds are also presented.

Figure 5.6. Model estimation of the difference in entire poleward heat transfer (in PW), by oceans and atmosphere between the modern-day and the Pliocene climate modeling experiment of (Haywood et al. (2000a) (xajpa = present-day simulation; xaiud = mid-Pliocene simulation).

Figure 6.1. Variations in foraminifera's isotopic composition through deep-sea cores show a gradual cooling of the global temperature. Greater negative $d^{18}O$ values characterize times of global warming and low polar ice volume throughout the Cretaceous and Eocene. In contrast, more + values through the Oligocene during Quaternary indicate steady cooling and the formation of polar ice caps.

Figure 6.2. For the last 1,800 years, tree rings, lake, and estuarine sediments, and arctic and alpine ice cores have been used to reconstruct and measure global mean temperatures. (a) Reconstructed temperatures combination Southern and Northern Hemisphere records. (b) Centered on the same instrumental and reconstructed temperature curves, hemispheric temperature difference (Mann and Jones, 2003).

Figure 6.3. For the past 100,000 years, GRIP ice core isotope record Greenland, Summit. On average, the oxygen isotope is 55 cm per mil.

Figure 6.4. For the past 400,000 years, the Nevada, Devils Hole, core DH-11 vein calcite oxygen-18 and the Vostok, Antarctica, ice core deuterium time series.

Figure 6.5. Temperature fluctuations at Summit, Greenland GRIP ice core reconstructed through oxygen isotopes, and central European temperatures reconstructed through Ammersee ostracode oxygen isotopes.

Figure 7.1. Variability in Holocene climate across time.

Figure 7.2. From the recent glacial maximum, the Northern Hemisphere's solar radiation and other climatic factors have changed. (Kutzbach and Street-Perrott, 1985).

Figure 7.3. A record of tropical African lake levels is correlated to the concentrations of atmospheric methane found in a Greenland core of ice (Roberts, 1998; Gillespie et al., 1983; Blunier et al., 1995). Lengthier climate patterns are interrupted by abrupt arid occurrences, as seen by the arrows. The times are in 1000.0 yBP (cal).

Figure 7.4. Correlation among macrofossil and pollen reports of great elevation glacier and places variations through the former 12,000 years; (A) percentages of pollen and Gouille´ Rion macrofossil concentrations (Swiss Alps); (B) percentages of pollen at Lengi Egga (Swiss Alps); (C) radiocarbon wood dates and debris of organic accumulated in Alpine glaciers front (Hormes et al., 2001); (D) Swiss glaciers determined length change (Maisch et al., 1999); (E) cold-humid phases of central European (Haas et al., 1998). +in a and b determine the chronological location of radiocarbon dates worked for depth-age models.

Figure 7.5. Depending on pollen sequences and radiocarbon-dated macrofossil, the estimated treeline elevation (m) and Holocene timberline in the central Alps of Swiss (Tinner and Theurillat, 2003).

Figure 7.6. Depending on wood macrofossils of radiocarbon-dated retrieve from high-elevation places, this is the approximate minimal earliest mid-Holocene hillside height in the Canadian Cordillera (southern) (Reasoner et al., 2001).

Figure 7.7. Calendar-dated snags discovered from the Site of Athabasca Glacier maintained the duration of the record. Fallen and Standing snags can be found near and over the current treeline, or if the snags are much larger than that of the trees that are currently growing in the area. The Athabasca Glacier's significant Small Ice Age advancements have been timed to 1714 (Luckman, 1994).

Figure 8.1. After the Pleistocene era's ice age, the course of climate, showcasing a typical culminating and cooling at the time of the Cenozoic age is outlined by the graph. Discovered in marine sediment documents, the oxygen isotope document of foraminifers helps interpret the comparative temperatures. On the right are the primate evolution's chief incidents. (Partridge et al., 1995; Tattersall, 1993).

Figure 8.2. Represented through the carbon isotopes in soils' interpretation of vegetation document, the Milankovitch cycles' supremacy (far right), oxygen isotopes from the marine document's interpretation of worldwide temperatures; human being phylogeny's graph of the assessment of the African climate inconsistency (Wood, 1992; Cerling, 1992; Cerling and Hay, 1988; deMenocal and Bloemendal, 1995; Shackleton et al., 1990).

Figure 8.3. Hominin evolution's outline.

Figure 8.4. Evolution and climate change.

Figure 8.5. Ice cliffs facing dry-seasonal dust bands have resulted from precipitation's diverse seasonality at places such as the ice cap of Quelccaya. Extremely precise dates of archives of ice cores are provided by those layers which makes them very valuable to paleo climatologists.

LIST OF TABLES

Table 6.1. Summary of methods employed to reconstruct atmospheric temperatures

LIST OF ABBREVIATIONS

ACC	Antarctic Circumpolar Current
AGCM	Atmospheric General Circulation Model
AOGCM	Atmosphere-Ocean General Circulation Model
BFE	Benthic Foraminiferal Extinction
CCSM	Community Climate System Model
CGCM	Coupled General Circulation Model
EAIS	East Antarctic Ice Sheet
EPICA	European Program for Ice Coring
FGOALS	Flexible Global Ocean-Atmosphere-Land System
GCMs	General Circulation Models
GHG	Greenhouse Gas
GIA	Glacial Isostatic Adjustment
GISP II	Greenland Ice Sheet Project II
GRIP	Greenland Ice Core Project
GSSP	Global Stratigraphic Section and Point
HadCM	Hadley Center Coupled Model
IETM	Initial Eocene Thermal Maximum
INQUA	International Union for Quaternary Research
IPSL-CM	Institut Pierre Simon Laplace Climate System Model
IRD	Ice-Rafted Detritus
JJA	June, July, and August
LGM	Last Glacial Maximum
LPTM	Late Paleocene Thermal Maximum
MAT	Modern Analog Technique
MIROC	Model for Interdisciplinary Research on Climate
MOC	Meridional Overturning Circulation
NGRIP	North Greenland Ice Core Project
NH	North Hemisphere

OGCM	Ocean General Circulation Model
OHT	Ocean Heat Transport
PETM	Paleocene Eocene Thermal Maximum
QBO	Quasi-Biennial Oscillation
RSL	Relative Sea Level
SH	South Hemisphere
SST	Sea Surface Temperature
TAR	Third Assessment Report
WSBW	Warm Saline Bottom Water

PREFACE

Since the birth of civilization, Earth Systems have undergone several climate changes. Understanding the fundamentals concepts concerning paleoclimates and climate change is much more need of time than ever before. This discussion would bring an enlightened approach towards the conceptualization of interacting components of the Earth system and their various features affecting the climate.

For centuries, literature on geology has laid much importance towards the deposition and extraction framework of the sediments found on the continents; however, the least importance has been given to the history of the past climates of Earth. Temperature and carbon cycles are considered the basis of the climatic phenomena in the history of climate change. Concepts of these climate cycles have been extensively elaborated in the discussion. A scientific strategy has been adopted to describe climatic variations occurring in the past. Solar radiations, events of the asteroid collision, endogenic heating, and plate tectonics have been some of the major causes of climate change. Other factors which might have influenced the overall structure of the Earth System include mountains terrain, oceanic circulation, solar, and lunar eclipses, volcanic eruptions, sand, and dust storms, along with the levels of the lake and fluvial signs. Paleoclimatology offers a great deal of discussion on past climate events for our understanding and application on predictable future events regarding climate change.

Moreover, emphasis has been given to the structure and chemical composition of the plants in the section of paleobotany. Basics of wood and leaf's structure remain of significant importance for the current developmental growth of plants. In addition to this, a paleocene model for investigation has been carried out to explain meridional overturning circulation (MOC) and deep-water formation. Paleoceanic techniques have been developed to gather data on stable isotopes, radiogenic isotopes, biogenic compounds and, elements. In the same way, paleoceanic processes such as paleotides, radiation, change in the oceanic base and, ice have been discussed in detail. Additionally, Paleocene-Eocene Thermal Maximum (PETM) modeling and methods of climate change have also been part of the discussion.

Pliocene climates and climate change is a debate of through discussion in chapter 5 of this book. Variation in sea levels, ice sheets, and sea ice and their declining trends in the past has a profound influence on the future climates have been thoroughly explained. Factors that have been affected the most are surface temperature, total precipitation rate, wind, and surface pressure, atmospheric CO2 and, oceanic heat transfer. Variation in the above-mentioned parameters has a drastic impact on the overall trends of climate change, as explained in the investigation models. Holocene climate is another major topic discussed in this book. Ancient Holocene climates, their variation trends and, their dynamics have been elaborated creating a link between the past and the future.

In the end, worldwide climate change and its impact on human evolution have been discussed, keeping in view the benefits offered by paleoclimatology. This book can deliver a great deal of vital information to scientists, environmentalists, policymakers, professors, and students. Students and teachers from the background of environmental science, geology, petrology, botany, and zoology can take away valuable information from this book.

Chapter 1

Fundamentals of Paleoclimatology and Climate Change

CONTENTS

1.1. Introduction..2
1.2. Paleoclimatology: Historical Climate 3
1.3. Paleoclimatic Methods...5
1.4. The Pre-Quaternary Climates 10
1.5. Glacial Interglacial Dynamics and Variability....................................15
References...32

1.1. INTRODUCTION

Climate is the average of the weather for an area over a period. There is a common misperception that arises when the public uses the term climate. Often, even if they thought they are debating climate, the conversation would begin with a discussion of the existing weather. It might go something like this:

"It was certainly hot today! Must be global warming…"

and it might be. Though, what they are telling is the weather. Weather is the summation of existing atmospheric conditions at any given period, and, to measure it, we utilize things like barometers, anemometers, thermometers, and other devices. This information is gathered and is utilized to develop models that could be used for analytical purposes. From there, you got your 1–10-day forecasts. These forecasts of the weather are completely founded on probabilities. A 20% probability of rain just means that, under current weather conditions (or that forecast), it tends to rain 20% of the time. These conditions are a portion of an area's climate story; however, they are still termed as the weather. Though they use the latest historical data, they are planned to explain the condition of the atmosphere at a moment in time… possibly three days from now (Figure 1.1).

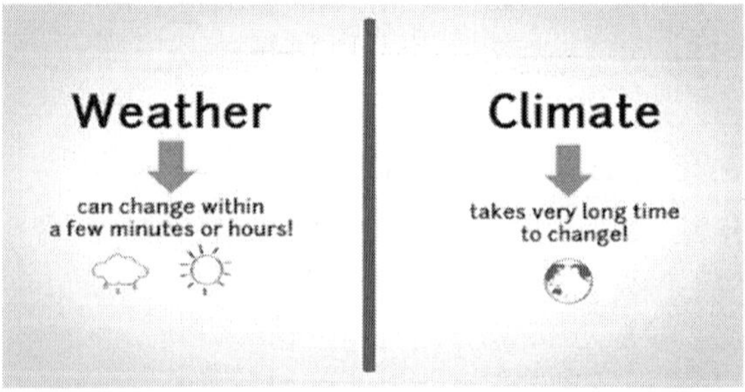

Figure 1.1. Illustration of weather versus climate.

Source: https://www.pinterest.com/pin/487092515940280512/.

Climate, on the other hand, is the average of these day-to-day conditions over time. At its most general and familiar level, the climate could be explained based upon seasons. Few places on Earth experience 4 seasons (winter through autumn), others experience 2 (rainy and dry). These short-term climatic situations are founded upon a large array of variables. These comprise topography, latitude, time of year, proximity to the sea, and so forth. Looking at variations in climate over the passage of year at a specific location on the planet gives a short-term picture distinctive to that location. Climate could also be defined on a global average basis. Under these circumstances, we could follow long-term variations in temperature and related variables (ice loss, evapotranspiration, greenhouse gases (GHGs), etc.), that gives a wider picture of global climate and how it is varying.

1.2. PALEOCLIMATOLOGY: HISTORICAL CLIMATE

Paleoclimatology is the study of the Earth's climate earlier to instrumental records were obtainable. The differentiation between climatology and paleoclimatology is similar to biology and paleobiology. Both fields study climate, however, one is very much attentive to our existing climatic situation. Paleoclimatologists must tease out the story of the climatic past of the land, that past is full of proof of change (Figure 1.2).

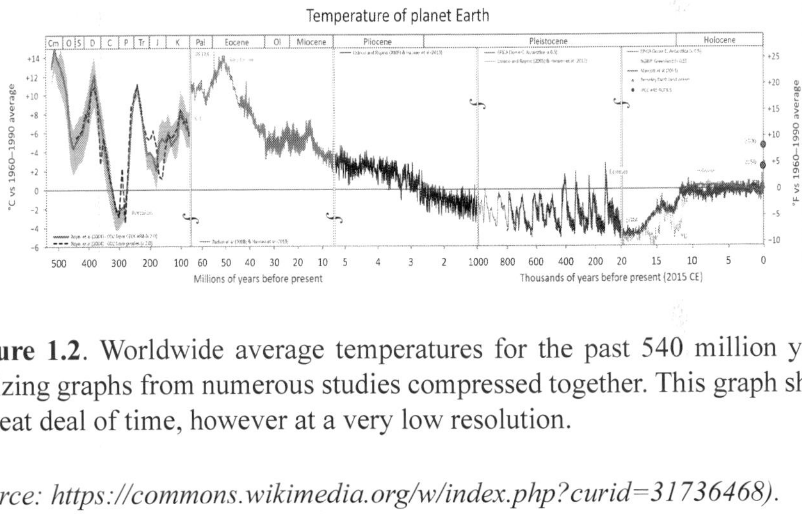

Figure 1.2. Worldwide average temperatures for the past 540 million years, utilizing graphs from numerous studies compressed together. This graph shows a great deal of time, however at a very low resolution.

Source: https://commons.wikimedia.org/w/index.php?curid=31736468).

While we incline to emphasize measuring our climate and weather, through utilizing instrumentation in space and on the ground, the Earth has kept a record of its previous climate in an extensive variety of ways. In this chapter, we would explore an extensive array of these types of paleoclimate evidence, comprising things like layers of ice, tree rings, and pollen. There are many more instances. Few of these records great detail over little periods however few records more rough detail over much larger periods. This means that the Earth's data stores are certainly no different than the data stores we produce in instrumentation. Measurement of climate is a matter of resolution and time. The utility of a specific data set is restricted by time and resolution. Hence, like studies of modern climate, paleoclimatologists ask queries and follow investigations that are directed by data restricted by time and resolution (Figure 1.3).

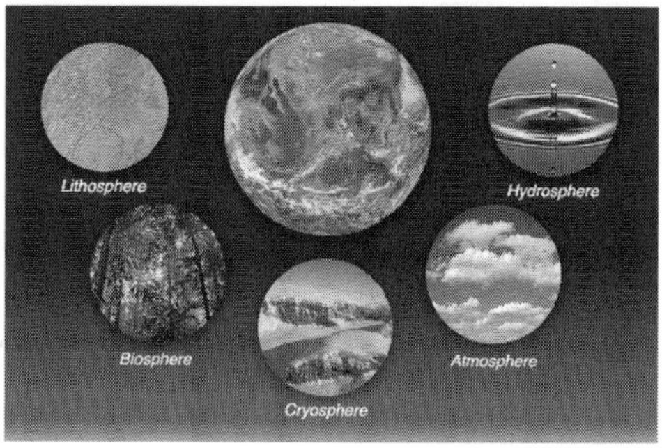

Figure 1.3. Earth's various systems.

Source: https://opengeology.org/historicalgeology/paleoclimatology-earth-systems-change-through-time/.

There are some significant assumptions recognized when studying paleoclimate. Like in all of the geosciences, these expectations are uniformitarian. That is, we assume previous processes normally operated the same method as in the present. That does not mean variations in the Earth's systems that disturb climate always happen similarly, however just that the processes that power or drive the changes to do (Figure 1.4).

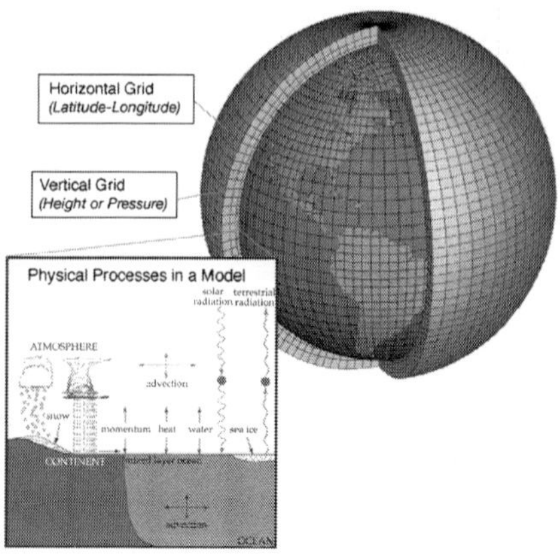

Figure 1.4. Illustration of climate models.

Source: https://opengeology.org/historicalgeology/paleoclimatology-earth-systems-change-through-time/.

Note: Climate models are formed up of systems of equations that try to account for an extensive variety of variables and their interfaces. Known past data is taken into account in making those equations and physical processes, few of which are recorded in the inset here, are the emphasis. To run the model on a supercomputer, the atmosphere and globe are divided into equations run and grid.

The study of paleoclimate has led to an extensive variety of insights into our recent climate and how it operates. It has assisted to improve the models we use to study existing problems related to anthropogenic (human-produced) warming. Studies on paleoclimate had led to a comprehensive knowledge of exactly how our solar system functions. Though sunspots were discovered by Galileo and others, the periodicity also the impact on climate at diverse magnitudes and scales would not be identified if the research is not for paleoclimatic. Similarly, we would possibly not know that the Earth, like other planets, sees steady variations in its orbital parameters that distress its climate. Knowledge about these exospheric effects informs our existing research, not only because it assists us to comprehend how our climate should be replying under only normal influences, but also because we could remove them as mechanisms besides the current variations we are producing. Paleoclimate research is critically significant to understanding our future.

The section evaluates data of paleoclimatic and information exactly how the climate system alters from interannual to millennial time scales, also how fine these changes be replicated through models of climate. Further paleoclimatic viewpoints are comprised in the next section (Rahmstorf, 2001; Alley et al., 2003; Overpeck and Trenberth, 2004).

1.3. PALEOCLIMATIC METHODS

1.3.1. Observations of Forcing and Response

The field of paleoclimatology had perceived substantial procedural progress since the TAR (third assessment report), also the drive of this part is to focus on these advances however providing an outline of the techniques underlying the data used in this section. Numerous crucial procedural details are accessible in the following sections where required.

Therefore, this procedures section is made to be more common and to provide readers more comprehension and assurance to find the chapter. Students are denoted to numerous valuable books and special matters of journals for further procedural detail (Bradley, 1999; Cronin, 1999; Fischer and Wefer, 1999; Ruddiman and Thomson, 2001; Alverson et al., 2003; Mackay et al., 2003; Kucera et al., 2005; NRC, 2006).

1.3.1.1. How Are Former Climate Forcing Recognized?

Time sequences of exorbitantly driven solation variation are well recognized also could be calculated from celestial mechanics. The techniques besides reconstructions of former solar and volcanic forcing remain to develop, though significant reservations still occur.

1.3.1.2. How Are Former Variations in Global Atmospheric Composition Recognized?

Possibly one of the greatest significant aspects of the latest paleoclimatology is that it is probable to derive time sequences of atmospheric trace aerosols and gases for the time from around 650 kyr to the existing from the ice itself and air trapped in polar ice. It is usual in paleoclimatic studies of the Late Quaternary, quality of response series and forcing are confirmed against the latest (for example, post-1950) measurements done through direct instrumental sampling. As is usual across all phases of the field, paleoclimatologists rarely depend on a single proxy or method, however rather on many. This gives a richer and more encompassing outline of climatic alteration than would be accessible from one proxy. Through this way, outcomes could be cross-checked and doubts could be understood. In the situation of pre-Quaternary CO_2, various biological and geochemical techniques give reasonable limitations on former CO_2 changes, however, the quality of the approximations is somewhat restricted.

1.3.1.3. How Exactly Could Paleoclimatic Records of Forcing and Response Be Dated?

Considerable had been researched and noted on the dating techniques related to paleoclimatic chronicles, and students are guided for more detail they might refer to the background books mentioned above. Generally, dating precision becomes feebler farther back in time and dating techniques normally had precise ranges where they could be used. Tree ring chronicles are usually the most precise also they are accurate to the season of a year or year (even earlier thousands of years). It had a host of further substitutes that also had yearly bands or layers (for example, varved sediments, some ice cores, corals, some cave deposits) however the age types related to these are not constantly precise for a particular annum. The Paleoclimatologists attempt to produce information of age through several sources to decrease ambiguity on age and paleoclimatic clarifications must be considered into account doubts in time controlling.

They continued for substantial developments in radiometric dating. Every system of radiometric had ranges through which the structure is beneficial, also studies of paleoclimatic nearly always print analytical doubts. As there could be additional doubts, techniques had been formed for verifying expectations and cross authenticating through independent techniques. For instance, differences of secular in the radiocarbon watch above the latter 12 kyr are well recognized, also properly well understood on the final 35 kyr. The differences, also the radiocarbon clock quality, both being well demonstrated through contrasts with age types resultant from varved sediment archives and specific tree rings, also with age determinations resultant from autonomous radiometric structures like series of uranium. Though, for every proxy chronicle, the quality of radiocarbon chronology also relies on the material accessible for dating density of dates and information regarding the radiocarbon phase of carbon that was integrated into the dated material.

1.3.1.4. How Could Paleoclimatic Proxy Techniques Be Used to Reconstruct Former Climate Dynamics?

Mostly the techniques regarding paleoclimatic reconstructions considered in this section are explained in certain detail in the above-mentioned books, also in the records of every chapter section. In certain units, significant methodological contextual also disagreements are debated where these deliberations assist to evaluate uncertainties of paleoclimatic.

Reconstruction techniques of paleoclimatic had developed significantly in the earlier periods, and extent from direct calculations of past variation (for example, ground temperature differences, glacier extent variations, gas contented in air bubbles of ice core, and ocean residue pore water variation) to substitute measurements comprising the alteration in physical, biological, and chemical parameters that imitate repeatedly in a well-understood and quantitative manner, earlier alteration in the surroundings where the proxy mover produced or occurred. Additionally, these techniques, paleoclimatologists also utilize documentary data (for example, in the type of precise logs, observations, and data of crop harvest) for reconstructions of former climates. However, many doubts remain, it is currently verified and well recognized that numerous organisms (for example, plankton insects, corals, trees, and other organisms) modify their development or/ and population changing aspects in relying on altering climate, and that the climate persuaded variations are completely noted in the former development of fossil (dead), also living specimens or organisms' accumulations. Lake plankton, pollen, ocean, and tree rings, are a few of the best recognized also well-formed proxy bases of former climate moving past centuries and ages.

The tree ring network density also thickness chronologies are utilized to gather earlier temperature and moisture variations founded on complete adjustment through temporary coinciding instrumental data. Former deliveries of plankton also pollen through the cores of sediment could be utilized to derive measurable approximations of former climate (for example, precipitation, temperatures, and salinity) through statistical techniques adjusted against the elated parameters of climate and their recent distribution.

The chemistry of numerous physical and biological entities reveals a complete understanding of thermodynamic procedures that could be converted into approximations of climate limitations like temperature. Main instances comprise O (oxygen) isotope ratios in the foraminiferal carbonate and coral to gather previous salinity and temperature; Mg/Ca (magnesium/ calcium) and Sr/Ca (strontium/ calcium) percentages in carbonate for the temperature estimation; saturation of alkenone taken from the oceanic organic molecules to gather former SST (sea surface temperature); hydrogen and Oxygen isotopes joined argon and nitrogen isotope studies in the cores of ice to gather atmospheric and temperature carriage.

Finally, numerous physical systems (for example, aeolian, and sediments deposits) alter in a predictable method that could be utilized to infer former climate alteration. There is continuing action on additional progress and improvement of techniques, and there are outstanding research problems regarding the extent to which the techniques had seasonal also spatial biases. Thus, in numerous latest paleoclimatic research, a grouping of techniques is used as multi-proxy series gives additional severe approximations than a one proxy method, and the multi-proxy method might recognize probable seasonal biases in the approximations. Not any paleoclimatic technique is perfect, and information of the essential techniques and procedures is needed when utilizing data of paleoclimatic.

1.3.2. Modeling of Paleoclimate

Models of Climate are utilized to simulate chapters of previous climate (for example, the former Glacial Maximum, sudden climate events, or the last interglacial period) to assist in understanding the mechanisms of previous climate variations. Models are significant for testing physical assumptions quantitatively, as the theory of Milankovitch. Models permit the relation of cause and impact in former climate variation should be examined. Models also assist to complete the space amongst worldwide and paleoclimate local scale, like as paleoclimatic info is repeatedly patchy, seasonal, and scarce. For instance, records of stretched ice core display strong association amongst the worldwide assorted gases methane and CO_2 and native temperature in Antarctica, however, the fundamental relations amongst these variables are greatest explored through the model's assistance. Emerging a computable understanding of processes is the most efficient method to learn from earlier climate for the forthcoming, as there are possibly no direct equivalents of the upcoming in the former (Figure 1.5).

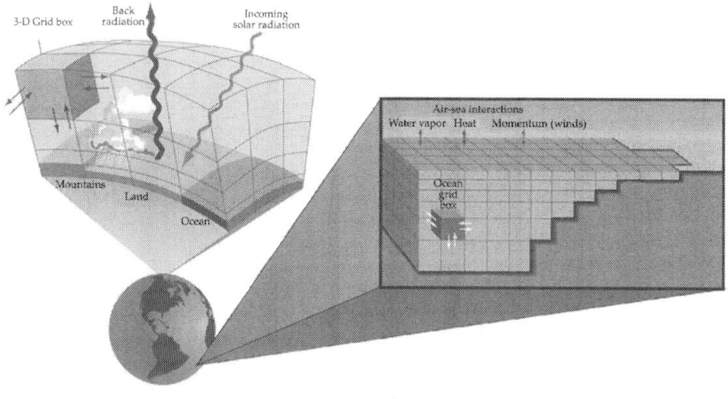

Figure 1.5. Diagram of the paleoclimate modeling.

Source: https://slideplayer.com/slide/13951470/.

At a similar period, paleoclimate reconstructions provide the probability for testing the climate models, especially if the forcing of climate could be properly stated, and the reply is appropriately well controlled. For past climates (for example, earlier to the existing 'Holocene' interglacial), compelling also replies cover a considerably greater extent, however, information is much scarce and indefinite, while for latest millennia more chronicles are accessible, however compelling also reply are very much lesser. Models' examination through paleoclimatic records is significant, however, all features of climate models could not be verified against instrumental climate data. For instance, a great presentation of the existing climate is not at all a decisive trial for an accurate carbon dioxide sensitivity to exam the same, simulation of climate through a very diverse carbon dioxide level could be utilized. Additionally, numerous parametrizations explaining the procedure of sub-grid measure (for example, mixing of turbulent, cloud parameters,) had been formed utilizing current day observations; therefore, states of climate that are not utilized in the development of the model give an independent standard for model testing. The data of Paleoclimate is significant for assessing the capability of climate models for simulation of accurate climate variation.

1.4. THE PRE-QUATERNARY CLIMATES

1.4.1. Association Between Temperature and Carbon Dioxide in This Period

The Pre-Quaternary climate earlier to 2.7 Ma was generally hotter than now and related to higher levels of carbon dioxide. In this situation, they had some resemblances with the expected upcoming climate variation (though worldwide geography and biology were progressively diverse further back in the period). Generally, they confirm that hotter climates are to be predicted through enhanced glasshouse gas absorptions. Observing previously in a period earlier to reaching of cores of ice, it is former to around 2 Ma, records on glasshouse gas absorptions in the troposphere develop very much indefinite. Though, there are continuing efforts to get measurable restorations of the hot climate above the previous 66 Myr, and the subsequent subsections deliberate two especially appropriate climate proceedings of this age.

Exactly how precisely is the association between temperature and carbon dioxide recognized? For the pre-Quaternary level of carbon dioxide, four key proxies are utilized (Hayes and Jasper, 1990; Royer, 2003; Royer et al., 2001). 2nd proxies relate to the point that entities of biological in seawater and soils had ratios of an isotope of carbon that are discrete from the troposphere (Cerling, 1991; Hayes and Freeman, 1992; Poths and Yapp, 1992; Pagani et al., 2005). 3rd proxy utilizes the ratio of boron isotopes (Palmer and Pearson, 2000), though the 4th utilizes the empirical connection between atmospheric carbon dioxide content and pores of stomatal on the tree leaves (Chaloner and McElwain, 1995; Royer, 2003). In Figure 1.6 it is revealed that (lowest panel), however, there is an extensive range of restored carbon dioxide values, magnitudes are usually greater than interglacial, values of pre-industrial observed in data of ice core. Variations in carbon dioxide on the extended time scales are assumed to be compelled through variations in tectonic procedures (for example, a drawdown of silicate weathering and source of volcanic activity; (Ruddiman, 1997)). Restorations of temperature, like that revealed in Figure 1.6 (central panel), are derivative from isotopes of oxygen (amended for changes in ice volume worldwide), also Magnesium/Calcium in alkenones and forms. Signs of the existence of continental ice on globe display that earth was generally free from ice throughout geologic history, one more sign of the overall hotness. The main extension of glaciations on the Antarctic beginning about 35–40 Ma was possibly a reply, in portion, to decreasing levels of atmospheric carbon dioxide from its highest in the Cretaceous ~100 Ma (Pollard and DeConto, 2003). The correlation amongst temperature and carbon dioxide could be found further back in the period like shown in Figure 1.6 (upper panel), which displays that the Mesozoic Age hotness (231 to 65 Ma) was probably related to huge levels of carbon dioxide and that the main glaciations around 300 Ma relatively accorded with less carbon dioxide concentrations comparative to surrounding times.

1.4.2. What Does the Mid Pliocene Record Show?

In the history of the world, the Mid Pliocene (around 3.4–3.1 Ma) is the most latest time which means that the temperature of the world was significantly hotter for a continued age (assessed through GCMs to be around 2–3°C over pre-industrial temperatures (Chandler et al., 1994; Sloan et al., 1996; Haywood et al., 2000; Jiang et al., 2005), giving an accessible instance of earth it is analogous in numerous forms to what estimation of models could be of the world of the previous 21st century. It is also the very latest that ocean also the continents basins had approximately reached the formation of their current geography. Collectively we could take, the hottest times average throughout the central Pliocene shows an observation of the steady-state of universally warmer earth, in which atmospheric carbon dioxide absorptions (projected to be among 360–400 ppm) were probably greater than pre-industrial standards (Rau and Raymo, 1992; Raymo et al., 1996), and a sign of isotopes and geologic decide that sea level was approximately 16–25 meter above recent levels (Cronin and Dowsett, 1990; Shackleton et al., 1995), with respectively lower continental dryness and decreased ice sheets (Guo et al., 2004).

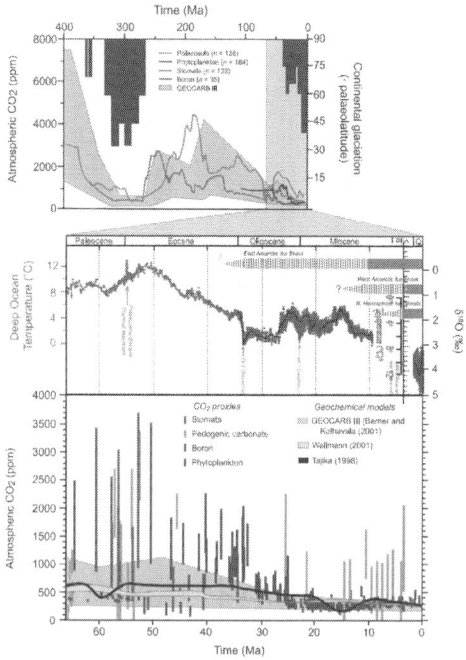

Figure 1.6. (Highest) Atmospheric carbon dioxide and continental glaciation of 400 Ma to current. (Central) Universal compiling of profound sea benthic foraminifera 180 isotope chronicles from 40 profound sea boring program and ocean boring program locations (Zachos et al., 2001) modernized through great resolution chronicles for the Eocene by Miocene break (Billups et al., 2002; Zachos and Bohaty, 2003; Lear et al., 2004). (Lowest) Comprehensive record of carbon dioxide for previous 65 Myr. Separate chronicles of carbon dioxide and related issues are color-coded through proxy technique; when probable, chronicles are founded on duplicate examples (Royer, 2006).

Source: https://www.semanticscholar.org/paper/Tectonic-boundary-conditions-for-climate-Crowley-Burke/cd979b6014e67ad34b32c5aaf2fe11d308852845.

Together marine and terrestrial paleoclimate substitutes (Thompson, 1991; Dowsett et al., 1996; Fleming and Thompson, 1996) displayed those high latitudes were considerably hotter, however, that humid surface air temperatures and SST were slightly different from the existing. This consequence is a significant reduction in the bottom tropospheric latitudinal hotness gradient. For instance, simulations of the atmospheric general circulation model (AGCM) are driven through restored SSTs from the Pliocene Research Interpretations and Synoptic Mapping Group (Dowsett et al., 1996, 2005) formed winter time temperature of surface air heating of 10 to 20°C at high north latitudes with 5 to 10°C rises above the northerly North Atlantic ~60° North, however, there was no variation of humid surface air temperature (or small cooling) (Chandler et al., 1994; Sloan et al., 1996; Haywood et al., 2000, Jiang et al., 2005). In comparison, a joined atmosphere-ocean experiment having an atmospheric carbon dioxide absorption of 400 ppm made heating comparative to pre-industrial times of 3 to 5°C in the northerly North Atlantic, and in the tropics, it is 1°C to 3°C (Haywood et al., 2005).

Estimated deficiency of tropical heating is a consequence of founding tropical SST restorations on the maritime microfaunal sign. As in the situation of the Final Glacial Maximum, it is indefinite whether the sensitivity of tropical is as slight as like reconstructions propose. (Haywood et al., 2005) noticed that approximates of alkenone of subtropical and tropical hotness do show heating in these areas, having a good agreement with general circulation model simulations from enhanced carbon dioxide forcing. The research recorded above, models of climate cannot form a reply to enhanced carbon dioxide through huge high latitude heating, also however the slight tropical temperature of tropical alter, without the tough rise in ocean warmth transport (Chandler and Rind, 1991).

1.4.3. What Does the Paleocene Eocene Thermal Record Extreme Display?

Nearly 56 Ma, sudden heating (in the situation of the instruction of 1–10 kyr) through numerous degrees Celsius is showed by fluctuating in oxygen isotope and Magnesium/Calcium chronicles (Stott and Kennett, 1991; Zachos et al., 2003; Tripati and Elderfield, 2004). The heating and related environmental effects were sensed at entire latitudes, both in the deep ocean and surface. The heat lasted nearly 101 kyr. Sign for moves in worldwide rainfall series is found in a diversity of fossil chronicles comprising flora (Wing et al., 2005). The irregularity of climate, alongside an associated carbon isotope trip, happened at the edges between the Eocene epochs and Paleocene and is hence regularly mentioned as the PETM (Paleocene Eocene thermal maximum). Thermal concentrated stands out in great resolution chronicles of that period. At a similar time, carbon isotopes in continental and marine chronicles display huge carbon mass having less concentration of carbon must have been discharged in the ocean also in atmosphere Mass of carbon was adequately huge to lessen the pH of the ocean and drive extensive suspension of the seafloor carbonates (Zachos et al., 2005). Probable bases for carbon could have been CH_4 from the clathrates decay on the sea ground, carbon dioxide, through sediments oxidation or volcanic activity, opulent in organic material (Dickens et al., 1997; Kurtz et al., 2003; Svensen et al., 2004). PETM, which changed ecosystems globally (Koch et al., 1992; Bowen et al., 2002, 2004; Bralower, 2002; Crouch et al., 2003; Thomas, 2003; Harrington et al., 2004), is intensively being deliberated as it had certain resemblance with the continuing quick discharge of carbon in the atmosphere through a human being. The projected carbon magnitude discharge for this date is of the phase of 1 to 2 × 1,018 gram of carbon (Dickens et al., 1997), and alike magnitude to that related to discharge of GHG throughout the upcoming century. Furthermore, the recovery period by natural carbon confiscation procedures, around 100 kyr, is alike to be predicted for the forthcoming. The situation of Pliocene, the high latitude heating throughout this happening was significant (~20°C; Moran et al., 2006) and extensively greater than formed through simulations of general circulation model for the occurrence (Sluijs et al., 2006) or usually for enhanced glasshouse gas tests. Though there is yet too much doubt in the records to derive a measurable approximation of climate compassion from the PETM, the occurrence is an outstanding instance of huge carbon discharge and associated extreme climatic heating (Figure 1.7).

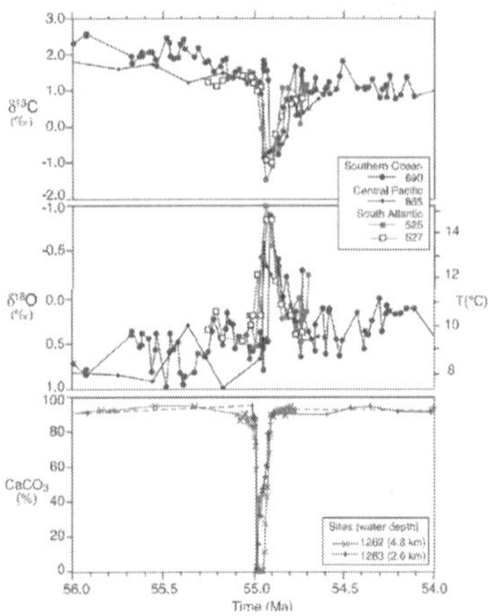

Figure 1.7. The PETM as chronicled in benthic (lowest dwelling) foraminifer (Nuttallides truempyi) isotopic chronicles from locations in the Pacific, South Atlantic, and Antarctic (Zachos et al., 2003).

Source: https://science.sciencemag.org/content/302/5650/1551.

Note: The quick reduction in ratios of an isotope of carbon in the upper section is symbolic of a huge rise in atmospheric glasshouse gases carbon dioxide also methane it was corresponding through around 5°C worldwide warming (mid panel). Utilizing the carbon isotope chronicles, arithmetical models display that methane released through the quick decay of maritime hydrates could have been the main constituent (~2,000 GtC) of the carbon fluctuation (Owen and Dickens, 1996). Trial of these models and additional ones need an autonomous limitation on the carbon fluctuations of carbon. The concept, many of the surplus glass house carbon might have been engrossed through the ocean, thus dropping the seawater pH of seawater, and producing extensive dissolution of sea ground carbonates. This reply is already obvious in the bottom section, which displays a transient decrease in $CaCO_3$ sediments content from the south Atlantic in 2 centers (Zachos et al., 2004, 2005). Perceived patterns show that the ocean's $CaCO_3$ ocean's saturation prospect quickly shoaled more than two kilometers, and later slowly improved as the ocean chemical balance was reinstated through the buffering procedures. Primarily, mostly $CaCO_3$ sediment dissolution is placed earlier to the happening, a procedure that balances the obvious time of the suspension horizon comparatively to the foundation of the benthic foraminifer isotope of carbon trip. Model Simulations display that retrieval of $CaCO_3$ saturation prospect must lead the retrieval in the elements of carbon as much as a 100 kyr (Owen and Dickens, 1996), an additional characteristic that is obvious in the sediment chronicles.

1.5. GLACIAL INTERGLACIAL DYNAMICS AND VARIABILITY

1.5.1. Forcing of Climate and Replying Over Glacial Interglacial Cycles

Chronicles of Paleoclimatic file a series of glacial-interglacial cycles layering the latter 740 kyr in cores of ice (EPICA Members of Community, 2004), also many million years in profound oceanic sediments of oceanic (Raymo and Lisiecki, 2005) also loess (Ding et al., 2002). The past 430 kyr, which are the finest recorded, are characterized through 100 kyr glacial-interglacial cycles of very huge amplitude, also as huge variations in climate conforming to further orbital ages (Hays et al., 1976), and at millenarian time scales (McManus et al., 2002; NorthGRIP, 2004). A minute quantity (20% on average) of every glacial-interglacial cycle was consumed in the method of a deep interglacial, which usually lasted for 10 to 30 kyr (Figure 1.8). It is an indication for the lengthier interglacial periods amongst 741 and 341 ka, however, these were seemingly cooler than the usual interglacial of the newest Quaternary (EPICA Members of Community, 2004). The newest interglacial Holocene spreads to the current.

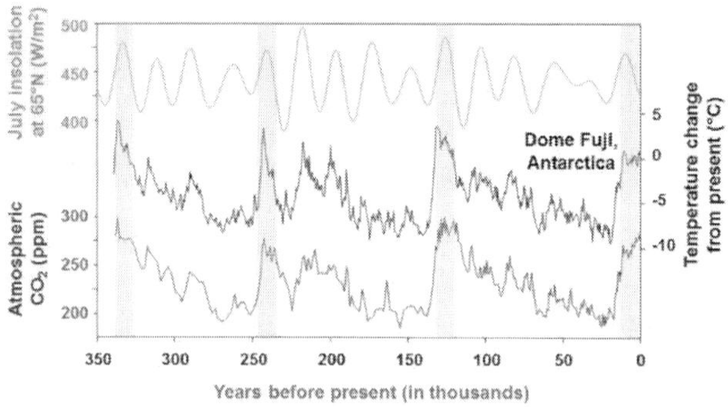

Figure 1.8. Record of glacial interglacial cycles.

Source: https://www.ncdc.noaa.gov/abrupt-climate-change/Glacial-Interglacial%20Cycles.

The core of ice data shows that glasshouse gases co diverse with the temperature of the Antarctic above glacial-interglacial cycles, proposing a close relation amongst temperature and usual atmospheric glasshouse gas alterations. Carbon dioxide variations over the past 400 kyr largely followed the temperature of the Antarctic, usually through several centuries to an epoch (Mudelsee, 2001). Series of climatic forcing and reactions throughout deglaciations (conversions from complete glacial situations to hot interglacial) are well recognized. Records of the high-resolution core of ice of temperature substitutes and carbon dioxide throughout deglaciation specify that temperature of Antarctic begins to increase many 100 years earlier to carbon dioxide (Monnin et al., 2001; Caillon et al., 2003). Throughout the latter deglaciation, and similarly also the 3 earlier, the start of heating at north and south latitudes led through many 1,000 annum the initial signs of substantial sea-level rise resultant through the north ice sheets melting associated through the fast heating at latitudes of upper north (Figure 1.9) (Petit et al., 1999; Shackleton, 2000; Pépin et al., 2001).

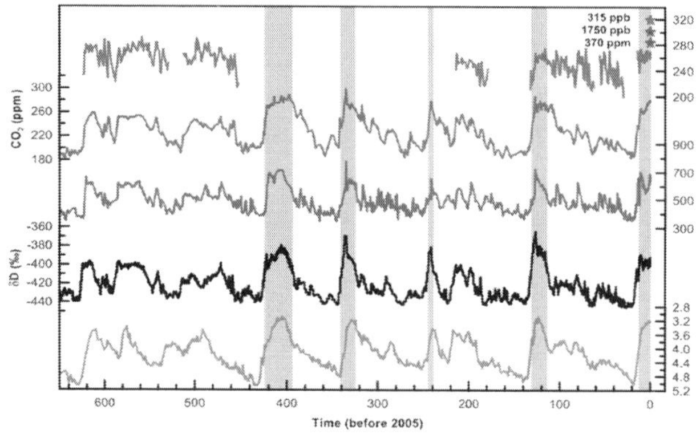

Figure 1.9. Differences of deuterium (δD; black), an indigenous temperature proxy, and the atmospheric absorptions of gases of the greenhouse carbon dioxide (red), methane (blue), and nitrous oxide (N_2O; green) taken from air stuck in cores of ice from Antarctica and latest measurements of atmospheric (Petit et al., 1999; Indermühle et al., 2000; EPICA Members of the Community, 2004; Spahni et al., 2005; Siegenthaler et al., 2005a, b).

Source: https://archive.ipcc.ch/publications_and_data/ar4/wg1/en/ch6.html.

Note: Shading shows the previous interglacial warm phases. The phases of interglacial also occurred earlier to 400 ka, however, these were seemingly cooler than the usual interglacial of the newest quaternary. The existing interglacial length is not rare in the situation of the previous 650 kyr. The mass of 57 universally dispersed benthic $\delta^{18}O$ chronicles of marine (dark grey), a substitution for universal ice volume variations (Raymo and Lisiecki, 2005), is presented for contrast with the data of ice core. The trends of descending in the benthic $\delta^{18}O$ curve reveal rising ice volumes on the surface. The shaded perpendicular bars are founded on the model of ice core age (EPICA Members of Community Members, 2004), and the record of marine is designed on its novel time scale founded on regulating the parameters of orbitals (Raymo and Lisiecki, 2005). The labels and stars show concentrations of atmosphere in 2000.

Existing data are not precise enough to find that warming initiated earlier in the NH (North Hemisphere) or SH (South Hemisphere), however, a main deglacial feature is the variance amongst South and North based on the timing and magnitude of strong reverses in the heating trend, which are not in stage amongst the hemispheres and are more marked in the NH (Brook and Blunier, 2001).

Carbon dioxide the GHG particularly feedbacks added significantly to the world radiative perturbation conforming to the conversions from glacial to interglacial means. Association between carbon dioxide and Antarctic temperature did not alter significantly throughout the last 650 kyr, showing rather a steady coupling amongst the carbon cycle and climate throughout the late Pleistocene (Siegenthaler et al., 2005a). The variation rate in atmospheric carbon dioxide varied significantly over time. For instance, the carbon dioxide rises from around 180 ppm at the latter glacial maximum to around 265 ppm in the initial Holocene happened with diverse rates over diverse times (Monnin et al., 2001).

1.5.1.1. How Do Glacial Interglacial Changes in the Greenhouse Gases (GHGs) Methane, Nitrous Oxide, and Carbon Dioxide Relate with the Industrialized Age Greenhouse Gas Rise?

Currently, atmospheric absorptions of carbon dioxide, methane, and nitrous oxide are greater than ever calculated in the core of ice data of the previous 650 kyr. The absorptions measurement of the three (3) GHGs altered only marginally (within 4% for N_2O and CO_2 and 7% for CH_4) over the previous millennium earlier to the industrialized era, and also changed within a limited range above the late quaternary. In the past 200 years, the late quaternary normal range had been surpassed through at least 25% for CO_2, 120% for CH_4, and 9% for N_2O. Records of all three display impacts of the huge and rising growth in anthropogenic discharges throughout the industrial era.

Changes in atmospheric carbon dioxide lead the radiative compelling through all 3 gases (Figure 1.9). The industrialized age increases the carbon dioxide, and in the radiative compelling through all 3 gases, is same in magnitude for the rise in the conversions from glacial to interglacial phases, however, initiated from an interglacial phase and happened 1 to 2 orders of quicker magnitude (Monnin and Stocker, 2003). There is no sign in the data of ice core that a rise similar in rate and magnitude to the industrialized age had happened in the former 650 kyr. Resolution of data is adequate to eliminate with very great confidence a peak same to the anthropogenic growth for the previous 50 kyr for carbon dioxide, for the previous 80 kyr for methane, and the previous 16 kyr for nitrous oxide. Records of the ice core demonstrate that throughout the industrialized age, the rise in the average rate of radiative forcing from carbon dioxide, methane, and nitrous oxide is higher than at any period during the previous 16 kyr (Figure 1.10). The atmospheric signal smoothing (Schwander et al., 1993; Spahni et al.,

2003) is little at Law Dome, a great accumulation location in Antarctica, and decadal gauge rates of variation could be calculated from the record of Law Dome crossing the previous 2 millennia (Etheridge et al., 1996; Ferretti et al., 2005; MacFarling et al., 2006). The rise in the average rate of atmospheric carbon dioxide was approximately 5 times greater in the period from 1960–1999 than in any 40 years throughout the two millennia earlier to the industrialized age. The rise in the average rate of atmospheric methane was approximately 6 times higher, and for nitrous oxide, approximately 2 times higher over the previous 4 decades, than at any period throughout the two millennia earlier to the industrialized age. Consistently, the latest rise in the average rate of the joined radiative forcing through all three GHGs was approximately 6 times higher than at any period throughout the period 1 to 1800 AD.

1.5.1.2. What Do the Former Glacial Maximum and the Former Deglaciation Show?

Previous glacial cold ages sometimes stated as 'ice ages,' give a means for assessing the understanding and showing the reply of the climate scheme to huge radiative alarms. Latest glacial age started around 116 ka, in reply to orbital forcing, by the rise of ice sheets and reduction of sea level concluding in the LGM (last glacial maximum), about 21 ka. The LGM and the following deglaciation had been extensively studied as the climate reply, radiative forcing, and boundary situations are comparatively well recognized.

Reply of the climate scheme at the LGM comprised responses on land and in the atmosphere increasing the orbital compelling. Absorptions of well merge GHGs at the LGM were decrease comparatively to pre-industrialized values (Figure 1.10), amounting to universal radiative consternation of –2.7 W m^{-2}-around equal to, however opposite from, the radiative compelling of these gases for year-end 2000. Huge parts of Europe and North America are ice sheltered at the LGM, dropping sea level and new land are exposed. The ice sheets radiative perturbation and dropped sea level, state as a boundary situation for certain LGM simulations, had been projected to be around –3.2 W m^{-2}, however with doubts related to the height and exposure of LGM continental ice (Mangerud et al., 2002; Peltier, 2004; Toracinta et al., 2004; Masson-Delmotte et al., 2006) and the ice parametrization albedo in models of climate (Taylor et al., 2000). The vegetation delivery was changed, with tundra extended above the northern continents and decrease tropical rain forest (Prentice et al., 2000), and atmospheric aerosols (mainly dust) were enhanced (Harrison and Kohfeld, 2001), partially as a result of decreased vegetation cover (Mahowald et al., 1999). Atmospheric aerosols and vegetation are taken as specified situations in few LGM simulations, every contributing around –1 W m^{-2} of radiative disconcertion, however with very little scientific knowledge of their radiative impact at the LGM (Claquin et al., 2003; Hewitt and Crucifix, 2005). Modifications in cycles of biogeochemical therefore played a significant role and added, through alterations in dust loading, GHG absorption, and vegetation cover, more than part of the recognized radiative perturbation throughout the LGM. Generally, the radiative disconcertion for the altered GHG, land surface, and concentrations of aerosol was around –8 W m^{-2} for the LGM, though with substantial uncertainty in the approximations for the assistances of aerosol and alterations in land surface (Figure 1.11).

Knowing of the magnitude of tropical coolness above the land at the LGM had upgraded as the TAR with additional chronicles, also as enhanced dating and climate signal understanding related with rising in snow line and vegetation alteration. Terrestrial climate restorations display tough spatial climatic data.

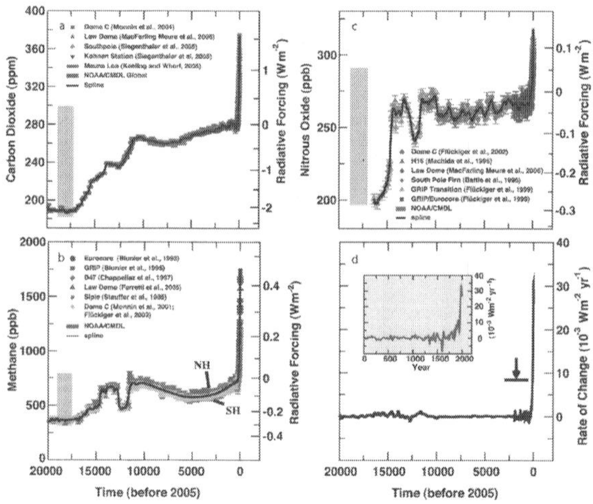

Figure 1.10. The radiative compelling and concentrations by (a) carbon dioxide; (b) methane; (c) nitrous oxide (N_2O); and (d) the alteration rate in their joined radiative forcing above the previous 20 kyr restored from Greenland and Antarctic ice and records (symbols) and measurements of direct atmosphere (magenta and red lines).

Source: https://archive.ipcc.ch/publications_and_data/ar4/wg1/en/ch6.html.

Note: Grey bars display the restored ranges of usual variability for the earlier 650 kyr (Siegenthaler et al., 2005a; Spahni et al., 2005). Radiative compelling was calculated with the basic terms of Chapter 2 (Myhre et al., 1998). The alteration rate in radiative compelling (black line) was calculated from spline fits (Enting, 1987) of the absorption data (black lines in boards a to c). The age dispersal thickness of the ice bubbles fluctuates from around 20 years for locations with a high gathering of snow like Antarctica, Law Dome, to around 200 years for little accumulation locations like Antarctica, Dome C. The firn data and Law Dome ice wrapping the previous two millennia, and latest instrumental data had been splined having 40 years cut off period, with the subsequent rate of alteration in radiative compelling shown through the inset in (d). Arrow indicates the topmost in the rate of alteration in radiative compelling later the anthropogenic signals of carbon dioxide, methane, and nitrous oxide had been smoothed through a model explaining the enclosure procedure of air stuck in ice (Spahni et al., 2003) applied for situations at the less accumulation Dome C location for the last glacial conversion. The carbon dioxide data are from Etheridge et al. (1996); Monnin et al. (2001, 2004); Siegenthaler et al. (2005a (Kohnen Station), b (South Pole)); and MacFarling Meure et al. (2006). The methane data are from Stauffer et al. (1985); Steele et al. (1992); Blunier et al. (1993, 1995); Dlugokencky et al. (1994); Chappellaz et al. (1997); Monnin et al. (2001); Flückiger et al. (2002); and Ferretti et al. (2005). The nitrous oxide data are from Machida et al. (1995); Battle et al. (1996); Flückiger et al. (1999, 2002); and MacFarling et al. (2006). Record of the atmosphere are from the National Oceanic and Atmospheric Administration's global air sampling network, demonstrating concentrations of universal average (dry air mole fraction; Steele et al., 1992; Dlugokencky et al., 1994; Conway, 2005), and by Mauna Loa, Hawaii (Whorf and Keeling, 2005). The record of universally averaged are accessible from http://www.cmdl.noaa.gov/.

World climate is confirmed through the balance of radiation of the earth. There are 3 essential methods the world's radiation balance could alter, thus causing climate change: (i) altering the arriving solar radiation (for example, by modifying in the World's orbit or in the Sun itself); (ii) altering the portion of solar radiation that is redirected (this portion is called the albedo) it could be altered, for instance by variations in cloud cover, little particles termed land cover or aerosols); and (iii) changing the extensive wave energy discharged back to space (for example, by alterations in GHG absorptions). Additionally, local climate also relies on how heat is dispersed through ocean and wind currents. All these factors had played a function in past climate variations. Initialing with the ice age that had come and gone in normal cycles from the previous approximately 3 million years ago, there is a strong indication that these are connected to regular changes in the World's orbit around the Sun, the so-termed Milankovitch cycles. These cycles alter the quantity of solar radiation reached at every latitude in every season (however hardly disturb the universal annual mean), and they could be measured with astronomical accuracy. There is yet some debate about how accurately these initiates and finishes ice ages, however many studies propose that the quantity of summer sunshine on north continents is critical: if it reduces below a serious value, snow from the previous winter does not melt in summer and an ice cover initiates to raise as more and more snow gathers. Simulations of the Climate model endorse that an Ice Age could certainly be initiated in this way, however, simple theoretical models had been utilized to effectively 'hindcast' the start of previous glaciations founded on the orbital variations. The subsequent huge decrease in north summer insolation, identical to those that initiated previous Ice Ages, is going to start in 30,000 years.

Though it is not the main reason, atmospheric carbon dioxide similarly plays a significant part in the ice ages. Record of the ice core of Antarctic display that carbon dioxide concentration is little in the icy glacial periods (~190 ppm), and large in the hot interglacial (~280 ppm); temperature alteration is followed by atmospheric carbon dioxide in Antarctica with an interval of many hundreds of years.

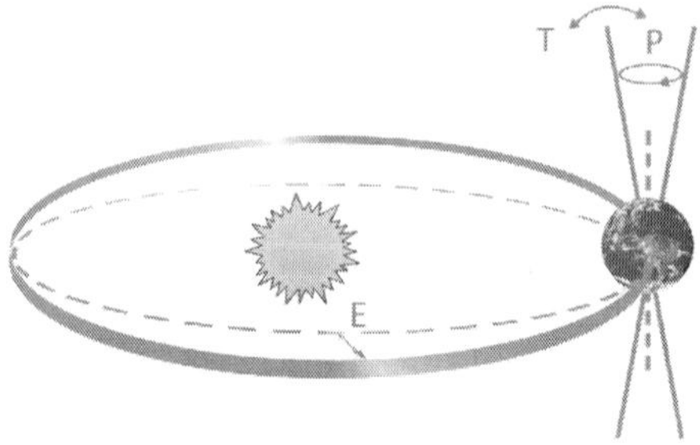

Figure 1.11. Diagram of the Earth's orbital variations (Milankovitch cycles) that led to the cycles of the ice age. 'T' means variations in the tilt (or obliquity) of the axis of Earth, 'E' means variations in the eccentricity of the orbit (because of differences in the slight axis of the ellipse), and 'P' means precession, that is, alterations in the path of the axis tilt at a specified point of the orbit (Schellnhuber and Rahmstorf, 2006).

Source: http://www.pik-potsdam.de/~stefan/klimawandel.html.

Due to the climate variations, at the start and finish of ice ages take many 1,000 years, many of these variations are affected through a positive carbon dioxide response; that is, a little early cooling because the Milankovitch cycles are successively enlarged as the carbon dioxide concentration decreases. Simulations of the model of ice age climate produce accurate results merely if the function of carbon dioxide is accounted for.

Throughout the latter ice age, more than 20 sudden and intense climate shifts happened that are especially prominent in chronicles around the north Atlantic. This fluctuation from the cycles of glacial-interglacial in that they possibly do not involve huge variations in the world mean temperature: variations are not synchronous in Antarctica and Greenland, and they are in the different direction in the North and South Atlantic. This signifies that the main difference in universal radiation balance might not have been required to cause these moves; a rearrangement of heat in the climate system would have served. Certainly, there is a strong indication that variations in ocean flow and heat transference could explain numerous features of these sudden events; model simulations and sediment data display that certain of these variations could have been activated through uncertainties in the ice sheets covering the Atlantic at the period, and the related freshwater discharge in the ocean.

1.5.1.3. How Results Are Accurate From Simulations of Climate Model of the Last Glacial Maximum (LGM)?

Intercomparisons of the model from the initial phase of the PMIP-1 (Paleoclimate Modelling Intercomparison Project), utilizing atmospheric models (rather with plain slab ocean models or with recommended SST), were presented in the TAR. Currently, there are 6 simulations of the LGM from the next phase (PMIP-2) utilizing AOGCMs (atmosphere-ocean general circulation models) and EMICs, though for this assessment only limited regional evaluations were finished in time. The radiative disconcertion for the PMIP-2 LGM simulations accessible for this valuation, which does not yet comprise the impacts of aerosol or vegetation variations, is –5 to –8 W m^{-2} (Figure 1.12).

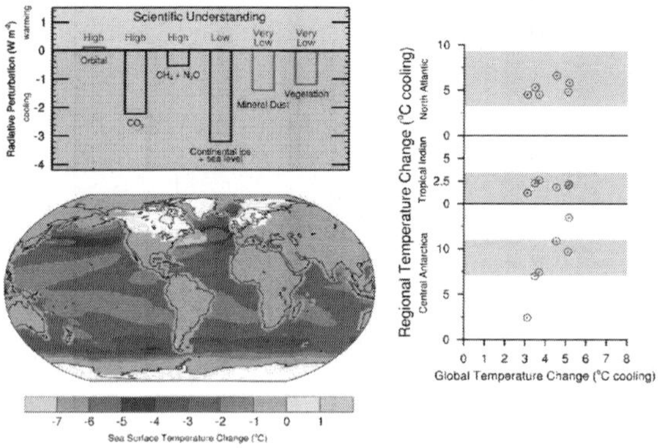

Figure 1.12. The LGM climate (nearly 21 ka) is comparative to the pre-industrialized (1750) climate.

Source: https://slideplayer.com/slide/15029904/.

Note: Global annual (Upper left) mean radiative impacts (W m^{-2}) of last glacial maximum climate variation agents,

usual responses in glacial-interglacial cycles, however, also stated in most AOGCM (atmosphere-ocean general circulation model) simulations for the last glacial maximum. The rectangular bar heights signify the best approximate values engaged by printed values of the climate alteration agents and transformation to radiative disconcertion utilizing simplified words for the greenhouse gas absorptions and calculations of model for the vegetation, mineral dust, and ice sheets. In-text references are comprised. Each approximation's dependability is a judgment by providing a level of scientific consideration founded on doubts in the climate variation agents and impacts of radiative physical understanding. PMIP-2 (Paleoclimate Modelling Intercomparison Project 2) simulations displayed in the lowest left and right panels don't comprise the radiative impacts of last glacial maximum alterations in vegetation or mineral dust (Lowest left) Multi-model average SST alter for last glacial maximum PMIP-2 imitations through 5 AOGCMs (HadCM (Hadley Center coupled model), MIROC (model for interdisciplinary research on climate), IPSL-CM (Institut Pierre Simon Laplace climate system model) FGOALS (flexible global ocean-atmosphere-land system), CCSM (community climate system model)). Ice-covered over continents is presented in white. (Right) last glacial maximum regional cooling matched to last glacial maximum global cooling as pretend in PMIP-2, with AOGCM outcomes presented as red circles and EMIC (ECBilt-CLIO) outcomes presented as blue circles. Regional averages are definite as Antarctica, yearly for internal ice cores; the tropical Indian Ocean, yearly for 16°S to 16°N, 51°E to 101°E; and North Atlantic Ocean, July-September for 41°N to 56°N, 36°W to 21°E. Grey shading shows the range of detected proxy approximations of regional chilling: Antarctica (Stenni et al., 2001; Masson-Delmotte et al., 2006), tropical Indian Ocean (Rosell-Mele et al., 2004; Barrows and Jug-Gins, 2005), and North Atlantic Ocean (Rosell-Mele et al., 2004; Kucera et al., 2005; Kageyama et al., 2006; de Vernal et al., 2006).

Paleoclimate Modelling Intercomparison Project-2 multi-model LGM SST alteration displays a modest chilling in the tropics, and highest chilling at latitudes of mid to high in relation with the rise in sea ice and variations in ocean movement (Figure 1.12). Paleoclimate Modelling Intercomparison Project-2 modeled firming of the SST meridional incline in the LGM North Atlantic, also as chilling and extended sea ice, confirms with substitution indicators (Kageyama et al., 2006). Polar extension of universal cooling, as noted in ice cores, is repeated for Antarctica (Figure 1.12), however the strong LGM chilling over Greenland is underrated, though with cautions regarding the altitudes of these ice caps in the Paleoclimate Modelling Intercomparison Project-2 imitations (Masson and Delmotte et al., 2006).

Paleoclimate Modelling Intercomparison Project-2 AOGCMs provide an extent of tropical ocean freezing amongst 16°S–16°N of 1.8°C–2.5°C. Sensitivity imitations with models show that this tropical freezing could be described through the decreased glacial GHG absorptions, which had direct impacts on the tropical radiative compelling (Shin et al., 2003; Otto-Bliesner et al., 2006b) and indirect impacts by LGM cooling through positive sea ice albedo response in the South Ocean backing to increase ventilation of ocean of the hot thermocline and the midway waters (Liu et al., 2002). Regional changes in pretended tropical chilling are much little than specified by MARGO data, partially linked to models at present resolutions being not capable to simulate the concentration of coastline surge and eastern border flows. Pretended to chill in the Indian Ocean (Figure 1.12), an important region with current day teleconnections to North Atlantic and Africa, links favorably to substitution approximations from alkenones (Rosell-Mele et al., 2004) and foraminifera accumulations (Juggins and Barrows, 2005).

Seeing variations in vegetation seems to expand the realism of imitations of the LGM, and facts to significant climate vegetation responses (McAvaney and Wyputta, 2001; Hewitt and Crucifix, 2005). For instance, expansion of the Asia tundra throughout the LGM adds to the cooling of the local surface, however, the heat of the tropics where savannah substitutes tropical forest (McAvaney and Wyputta, 2001). Responses between vegetation and climate happen locally, with a reduction in the tree portion in central Africa decreasing precipitation, and distantly with chilling in Siberia (tundra substituting trees) changing (lessening) the summer monsoon of Asia. The physiological influence of carbon dioxide absorption on vegetation required to be comprised to appropriately represent variations in the worldwide forest (Prentice and Harrison, 2003), also to broaden the climatic extent where shrubs and grasses lead. The biome spreading pretends with energetic world vegetation models repeats the extensive features witnessed in paleodata (for example, Prentice and Harrison, 2003).

1.5.1.4. How Accurate Are Simulations of Earthly Carbon Storing at the Last Glacial Maximum (LGM)?

There is an indication that earthly carbon-storing was decreased throughout the LGM compared to now. Calculations on mass balance founded on measurements of carbon on covers of benthic foraminifera produce a decrease in the earthly biosphere carbon list (living and soil vegetation) of around 300–700 GtC (Shackleton, 1977; Bird et al., 1994) comparative to the pre-industrialized inventory of around 3,000 GtC. Approximations of terrestrial carbon-storing founded on reconstructions of the ecosystem propose an even bigger difference (for example, Crowley, 1995). Imitations with models of carbon cycle produce a decrease in world terrestrial carbon shares of 600–1,000 GtC at the LGM compared to the pre-industrialized period (Francois et al., 1998, 1999; Beerling, 1999; Kaplan et al., 2002, 2003; Liu et al., 2002; Joos et al., 2004). The majority of this imitation change is because of decreased simulated development causing by lesser atmospheric carbon dioxide. A main regulating function for carbon dioxide is consistent with the examination of model data of Bond et al. (2003), who proposed that little atmospheric carbon dioxide could have been a substantial factor in the decrease of trees throughout glacial times, due to their sluggish regrowth after turbulences like as fire. Consequences of terrestrial models, also utilized to project forthcoming carbon dioxide concentrations, are widely matched with the extent of reconstructed changes in glacial-interglacial carbon-storing on land.

1.5.1.5. How Lengthy Did the Earlier Interglacials Last?

Four interglacial of the former 450 kyr earlier the Holocene (Oceanic Isotope Phases 5, 7, 9, and 11) were all diverse in several features, comprising period (Figure 1.9). Shortest (Stage 7) took a few 1,000 centuries, the extended (Stage 11; ~420 to 395 ka) remained nearly 30 kyr. Indication for an abnormally extended Stage 11 had been recently secure through novel ice core also oceanic sediment data. European Program for Ice Coring (EPICA) in Antarctica records of Dome C Antarctic cores of ice proposes that temperature of Antarctic continued around as hot as the 28 kyr Holocene (EPICA Members of Community, 2004). A novel pile of 57 worldwide dispersed benthic $\delta^{18}O$ chronicles presents age approximations at Stage 11 closely similar to those given through the EPICA outcomes (Raymo and Lisiecki, 2005).

1.5.2. Sea Level Differences Over the Former Glacial-Interglacial Cycle

1.5.2.1. What Is the Impact of Past Ice Volume Alteration on Modern Sea Level Variation?

Previous chronicles of sea-level history give a critical basis for accepting the background differences on which where the sea level growth associated with modern procedures is overlaid. However, if no anthropogenic impact were presently functioning in the climate structure, computable, and substantial variations in RSL (relative sea level) would still be happening. The main reason for this natural inconsistency in sea level had to be done with the earth's memory of the former deglaciation incident. Although the so-termed GIA (glacial isostatic adjustment) procedure, equilibrium of gravitation is reinstated through deglaciation, not merely through crustal 'rebound,' however also by the horizontal redeployment of water in the ocean sinks needed to retain gravitational equipotential at the surface of the ocean.

Worldwide GIA procedure models had enabled the separation of an input to the latest rate of worldwide sea-level increase being calculated through the TOPEX (Topography Experiment) /Poseidon (T/P) satellite having –0.28-millimeter yr^{-1} for ICE-4G(VM2) Peltier model (1996) and –0.36-millimeter yr^{-1} for ICE-5G(VM2) (Peltier Model, 2004). Evaluations of this (Peltier, 2001) suggest that the influence of recent climate variation on the worldwide rate of sea-level increase is higher than implicit through the calculations of uncorrected T/P.

1.5.2.2. What was the Scale of Glacial-Interglacial Sea Level Alteration?

Evaluation of model-founded paleo sea level also assists to refine approximations of the worldwide averaged (eustatic) sea-level increase that happened throughout the latest glacial-interglacial from the LGM to the Holocene move. Stretched coral founded RSL curve from the Barbados island in the Caribbean Sea (Fairbanks, 1989; Fairbanks and Peltier, 2006) is particularly significant since the history of RSL from this location had been revealed to give a good estimate to the ice itself by comparable globally averaged curve (Peltier, 2002). Fitting of the estimate of the ICE 5G(VM2) model to the data of Fairbanks set, as revealed in Figure 1.13(B), restrains the remaining ice equal eustatic increase following to 21 ka to many 118.7 meters, very near to the value of around 120 meters conventionally contingent (for example, Shackleton, 2000) on the base of profound sea regarding data of Oxygen isotopic (Figure 1.13(B)). A reconstruction sea level formed by Waelbroeck et al. (2002) was founded upon chronicles of coral and profound sea Oxygen isotopes modified for the impact of the abyssal temperature of the ocean varies for the complete cycle of the glacial-interglacial. Record of same is characterized through the best approximation of the Last Greatest Maximum depression of ice comparable eustatic level of the sea that is too 120-meter close. Examination of the Red Sea Oxygen isotopic data through Siddal et al. (2003) more assist the legitimacy of the clarification of the comprehensive Barbados record through Fairbanks and Peltier (2006).

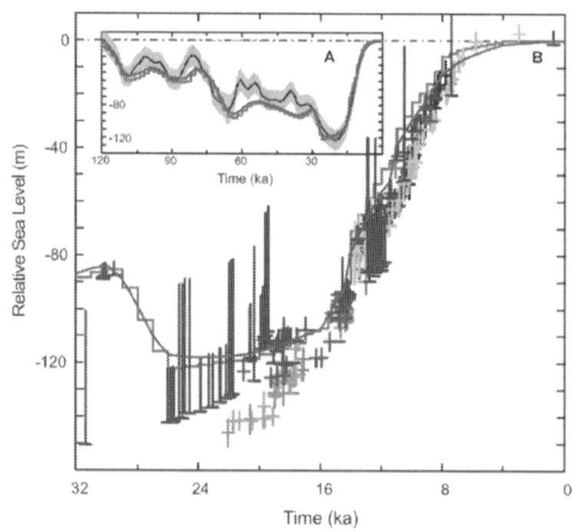

Figure 1.13. (A) Ice comparable of eustatic sea-level past above the cycle of the previous glacial-interglacial as per the Waelbroeck et al. (2002) examina-

tion. (B) Fitting of the ICE-5G(VM2) model forecast (red line) to the extensive inclined coral founded record of RSL history from the Barbados island in the Caribbean Sea (Fairbanks, 1989; Fairbanks and Peltier, 2006) above the age extent from 32 ka to current.

Source: https://www.sciencedirect.com/science/article/abs/pii/S0277379101001019.

Note: (A) Smooth black line describes the middle point of their approximations for every age and the nearby highlighted region gives an approximation of error. The Redline is the forecast of the ICE-5G(VM2) model for the location of Barbados on which the observations of RSL themselves provide an outstanding estimate to the ice comparable curve of eustatic sea level; (B) Real ice comparable eustatic curve of sea level for this model is presented as the step left brown line. Separate coral founded approximations of RSL (blue) had a fasten error bar that relies on the coral classes. Evaluations meant through the little error bars are taken from the species of Acropora palmitate, which gives the tightest restraints upon comparative sea level since this species is noticed to live in nearly 5 meters of sea level in the recent ecology. Evaluations meant by the lengthier error bars are taken rather from the species of coral Montastrea annularis (error bars of intermediary 20-meter length) or from other species that are initiated over an extensive range of bases with regarding the level of sea (lengthiest error bars). The supplementary information is most beneficial in giving a lesser bound in dejection of the sea level. Information signified through the colored marks are from the ice comparable eustatic reconstruction of sea level of Chappell and Lambeck (2001) for Huon (black), Sunda Shelf (magenta), Tahiti (grey), Barbados (cyan), and Bonaparte Gulf (orange).

1.5.2.3. What Is the Importance of Higher Than Current Sea Levels Throughout the Latter Interglacial Period?

Chronicles of eustatic sea-level alteration could be prolonged in the LIG time. Measurement of sea level directly founded upon coastline sedimentary deposits and sequences of tropical coral (for example, in tectonically steady settings) had recognized that eustatic sea level was larger than current throughout this latter interglacial through around four to six-meter (for example, Rostami et al., 2000; Muhs et al., 2002). Uninterrupted ice core chronicles of the NGRIP (North Greenland Ice Core Project) to 123 ka, older however distressed LIG ice in the GRIP (Greenland Ice Core Project) and GISP2 (Greenland Ice Sheet Project 2) cores, specify that the region of Greenland Summit stayed ice-covered throughout the LIG (Raynaud et al., 1997; NGRIP, 2004). Identical isotopic value changes found in the Renland cores and Camp Century (Johnsen et al., 2001) propose that comparative elevation changes throughout the LIG in north Greenland were not huge (NGRIP, 2004). Understanding of the Dye-3 core of ice in south Greenland is ambiguous. The existence of isotopically enhanced ice, probably LIG ice, on the lowest of the Dye-3 core had been taken as a considerable decrease in ice width of south Greenland throughout the LIG (NGRIP, 2004). Similar reasonable interpretations propose that the Ice Sheets of the south dome of Greenland did not live the highest interglacial heat and Dye-3 is noting the development of late LIG ice once the sheet of ice recognized itself in south Greenland (Fisher and Koerner, 2002), or ice that ran in the region from mid-Greenland or through a living however south dome are remote (Lhomme et al., 2005a). Lack of pre-LIG ice in the huge ice caps in the Canadian eastern Arctic specifies that throughout the LIG, they are entirely melted (Koerner, 1989).

Worldwide sea level mostly increases throughout the LIG must have been the outcome of melting polar ice sheet. Ice Sheet models of Greenland compulsory with the temperature of Greenland situations taken from data (Marshall and Cuffey, 2000; Peltier and Tarasov, 2003; Lhomme et al., 2005a), or temperatures and rainfall formed by an AOGCM (Otto-Bliesner et al., 2006a),1 pretend the nominal LIG Greenland Ice Sheet as a sharp sided, ice sheet in mid and north Greenland (Figure 1.11). Contingently ice sheets joined with the variation in further arctic ice fields, probably produced no more than 2–4 meters of initial LIG sea-level increase over many millennia. The imitation contribution of Greenland for this sea-level increase was probably driven through orbitally forced warming in the Arctic in summer.

1.5.2.4. What Is the Lengthy Term Input of Polar Ice-Sheet Resulting Melt Water to Detect Worldwide Averaged Rate of Sea Level Increase?

Postglacial RSL models of history organized with observations of Holocene could be used to evaluate rather or not an important fraction of the saw worldwide averaged rate of sea-level increase of around 2-millimeter yr^{-1} throughout the 20th century could be described as a lengthy-term lasting impact of the latest fractional deglaciation of the polar sheets of ice. Founded on the post-TAR evaluations resultant from interpretations of geological of Holocene sea-level through 16 equatorial Pacific islands (Peltier, 2002; Peltier et al., 2002), it seems probable that the increase of the average rate of sea-level because of this hypothetical basis over the previous 2 kyr was 0 and mostly in the extent from 0 to 0.2-millimeter yr^{-1} (Lambeck, 2002).

REFERENCES

1. Alley, R. B., et al., (2003). Abrupt climate change. *Science*, *299*(5615), 2005–2010.
2. Alverson, K. D., Bradley, R. S., & Pedersen, T. F., (2003). *Paleoclimate, Global Change, and the Future* (Vol. 1, p. 221). International geosphere biosphere program book series, Springer-Verlag, Berlin.
3. Barrows, T. T., & Juggins, S., (2005). Sea-surface temperatures around the Australian margin and Indian Ocean during the last glacial maximum. *Quat. Sci. Rev.*, *24*, 1017–1047.
4. Battle, M., et al., (1996). Atmospheric gas concentrations over the past century measured in air from firn at the South Pole. *Nature*, *383*(6597), 231–235.
5. Beerling, D. J., (1999). New estimates of carbon transfer to terrestrial ecosystems between the last glacial maximum and the Holocene. *Terra Nova*, *11*(4), 162–167.
6. Billups, K., Channell, J. E. T., & Zachos, J., (2002). Late Oligocene to early Miocene geochronology and paleoceanography from the subantarctic South Atlantic. *Paleoceanography*, *17*(1), 1004, doi: 10.1029/(2000)PA000568.
7. Bird, M. I., Lloyd, J., & Farquhar, G. D., (1994). Terrestrial carbon storage at the LGM. *Nature*, *371*(6498), 566.
8. Blunier, T., & Brook, E. J., (2001). Timing of millennial-scale climate change in Antarctica and Greenland during the last glacial period. *Science*, *291*, 109–112.
9. Blunier, T., et al., (1993). Atmospheric methane record from a Greenland ice core over the last 1,000 years. *Geophys. Res. Lett.*, *20*(20), 2219–2222.
10. Blunier, T., et al., (1995). Variations in atmospheric methane concentration during the Holocene epoch. *Nature*, *374*(6517), 46–49.
11. Bohaty, S. M., & Zachos, J. C., (2003). Significant Southern Ocean warming event in the late middle Eocene. *Geology*, *31*(11), 1017–1020.
12. Bond, W. J., Midgley, G. F., & Woodward, F. I., (2003). The importance of low atmospheric CO_2 and fire in promoting the spread of grasslands and savannas. *Global Change Biol.*, *9*(7), 973–982.
13. Bowen, G. J., et al., (2002). Mammalian dispersal at the Paleocene/Eocene boundary. *Science*, *295*(5562), 2062–2065.

14. Bowen, G. J., et al., (2004). A humid climate state during the Paleocene/ Eocene thermal maximum. *Nature*, *432*(7016), 495–499.
15. Bradley, R. S., (1999). Climatic variability in sixteenth-century Europe and its social dimension: Preface. *Clim. Change*, *43*(1), 1–2.
16. Bralower, T. J., (2002). Evidence of surface water oligotrophy during the Paleocene-Eocene thermal maximum: Nannofossil assemblage data from Ocean Drilling Program Site 690, Maud Rise, Weddell Sea. *Paleoceanography*, *17*(2), 1023, doi: 10.1029/(2001)PA000662.
17. Caillon, N., et al., (2003). Timing of atmospheric CO_2 and Antarctic temperature changes across termination III. *Science*, *299*, 1728–1731.
18. Cerling, T. E., (1991). Carbon dioxide in the atmosphere: Evidence from Cenozoic and Mesozoic paleosols. *Am. J. Sci.*, *291*, 377–400.
19. Chandler, M. A., Rind, D., & Thompson, R. S., (1994). Joint investigations of the middle Pliocene climate II: GISS GCM Northern hemisphere results. *Global Planet. Change*, *9*, 197–219.
20. Chappellaz, J., et al., (1997). Changes in the atmospheric CH_4 gradient between Greenland and Antarctica during the Holocene. *J. Geophys. Res.*, *102*(D13), 15987–15997.
21. Claquin, T., et al., (2003). Radiative forcing of climate by ice-age atmospheric dust. *Clim. Dyn.*, *20*, 193–202.
22. Cronin, T. M., (1999). *Principles of Paleoclimatology* (Vol. 1, p. 560). Perspectives in Paleobiology and Earth History. Columbia University Press, New York, NY.
23. Crouch, E. M., et al., (2003). The Apectodinium acme and terrestrial discharge during the Paleocene-Eocene thermal maximum: New palynological, geochemical and calcareous nannoplankton observations at Tawanui, New Zealand. *Paleogeogr. Paleoclimatol. Paleoecol*, *194*(4), 387–403.
24. Crowley, T. J., (1995). Ice-age terrestrial carbon changes revisited. *Global Biogeochem. Cycles*, *9*(3), 377–389.
25. Crucifix, M., & Hewitt, C. D., (2005). Impact of vegetation changes on the dynamics of the atmosphere at the last glacial maximum. *Clim. Dyn.*, *25*(5), 447–459.
26. Cuffey, K. M., & Marshall, S. J., (2000). Substantial contribution to sea-level rise during the last interglacial from the Greenland ice sheet. *Nature*, *404*, 591–594.

27. de Vernal, A., et al., (2006). Comparing proxies for the reconstruction of LGM sea-surface conditions in the northern North Atlantic. *Quat. Sci. Rev.*, *25*(21–22), 2820–2834.

28. DeConto, R. M., & Pollard, D., (2003). Rapid Cenozoic glaciation of Antarctica induced by declining atmospheric CO_2. *Nature*, *421*(6920), 245–249.

29. Dickens, G. R., & Owen, R. M., (1996). Sediment geochemical evidence for an early-middle Gilbert (early Pliocene) productivity peak in the North Pacific Red Clay Province. *Mar. Micropaleontol.*, *27*(1–4), 107–120.

30. Dickens, G. R., Castillo, M. M., & Walker, J. C. G., (1997). A blast of gas in the latest Paleocene: Simulating first-order effects of massive dissociation of oceanic methane hydrate. *Geology*, *25*(3), 259–262.

31. Ding, Z. L., et al., (2002). Stacked 2.6-Ma grain size record from the Chinese loess based on five sections and correlation with the deep-sea d18O record. *Paleoceanography*, *17*(3), 1033, doi: 10.1029/(2001) PA000725.

32. Dlugokencky, E. J., Steele, L. P., Lang, P. M., & Masarie, K. A., (1994). The growth rate and distribution of atmospheric methane. *J. Geophys. Res.*, *99*, 17021–17043.

33. Dowsett, H. J., & Cronin, T. M., (1990). High eustatic sea level during the middle Pliocene: Evidence from southeastern U. S. Atlantic coastal plain. *Geology*, *18*, 435–438.

34. Dowsett, H. J., Chandler, M. A., Cronin, T. M., & Dwyer, G. S., (2005). Middle Pliocene Sea surface temperature variability. *Paleoceanography*, *20*(2), doi: 10.1029/(2005)PA001133.

35. Dowsett, H., Barron, J., & Poore, R., (1996). Middle Pliocene Sea surface temperatures: A global reconstruction. *Mar. Micropaleontol.*, *27*(1–4), 13–25.

36. Enting, I. G., (1987). On the use of smoothing splines to filter CO_2 data. *J. Geophys. Res.*, *92*, 10977–10984.

37. EPICA Community Members, (2004). Eight glacial cycles from an Antarctic ice core. *Nature*, *429*(6992), 623–628.

38. Etheridge, D. M., et al., (1996). Natural and anthropogenic changes in atmospheric CO_2 over the last 1,000 years from air in Antarctic ice and firn. *J. Geophys. Res.*, *101*(D2), 4115–4128.

39. Fairbanks, R. G., (1989). A 17,000-year glacio-eustatic sea level record: Influence of glacial melting rates on the Younger Dryas event and deep- ocean circulation. *Paleoceanography*, *342*, 637–642.

40. Ferretti, D. F., et al., (2005). Unexpected changes to the global methane budget over the past (2000) years. *Science*, *309*, 1714–1717.

41. Fischer, G., & Wefer G., (1999). *Use of Proxies in Paleoceanography: Examples from the South Atlantic* (Vol. 1, p. 735). Springer, Berlin.

42. Flückiger, J., et al., (1999). Variations in atmospheric N_2O concentration during abrupt climatic changes. *Science*, *285*(5425), 227–230.

43. Flückiger, J., et al., (2002). High resolution Holocene N_2O ice core record and its relationship with CH_4 and CO_2. *Global Biogeochem. Cycles*, *16*, 1–20. doi: 10.1029/(2001)GB001417.

44. Francois, L. M., Delire, C., Warnant, P., & Munhoven, G., (1998). Modelling the glacial-interglacial changes in the continental biosphere. *Global Planet. Change*, *17*, 37–52.

45. Francois, L. M., et al., (1999). Carbon stocks and isotopic budgets of the terrestrial biosphere at mid-Holocene and last glacial maximum times. *Chem. Geol.*, *159*, 163–189.

46. Freeman, K. H., & Hayes, J. M., (1992). Fractionation of carbon isotopes by phytoplankton and estimates of ancient CO2 levels. *Global Biogeochem. Cycles*, *6*, 185–198.

47. Guo, Z. T., et al., (2004). Late Miocene-Pliocene development of Asian aridification as recorded in the Red-Earth formation in northern China. *Global Planet. Change*, *41*(3–4), 135–145.

48. Harrington, G. J., Kemp, S. J., & Koch, P. L., (2004). Paleocene-Eocene paratropical floral change in North America: Responses to climate change and plant immigration. *J. Geol. Soc. London*, *161*, 173–184.

49. Harrison, S. P., & Prentice, I. C., (2003). Climate and CO_2 controls on global vegetation distribution at the last glacial maximum: Analysis based on paleovegetation data, biome modelling and paleoclimate simulations. *Global Change Biol.*, *9*, 983–1004.

50. Haywood, A. M., Dekens, P., Ravelo, A. C., & Williams, M., (2005). Warmer tropics during the mid-Pliocene? Evidence from alkenone paleothermometry and a fully coupled ocean-atmosphere GCM. *Geochem. Geophys. Geosystems*, *6*, Q03010, doi: 10.1029/(2004) GC000799.

51. Haywood, A. M., Valdes, P. J., & Sellwood, B. W., (2000). Global scale paleoclimate reconstruction of the middle Pliocene climate using the UKMO GCM: Initial results. *Global Planet. Change*, *25*, 239–256.
52. Indermühle, A., et al., (2000). Atmospheric CO_2 concentration from 60 to 20 kyr BP from the Taylor Dome ice core, Antarctica. *Geophys. Res. Lett.*, *27*(5), 735–738.
53. Jasper, J. P., & Hayes, J. M., (1990). A carbon isotope record of CO_2 levels during the late Quaternary. *Nature*, *347*, 462–464.
54. Jiang, D., et al., (2005). Modeling the middle Pliocene climate with a global atmospheric general circulation model. *J. Geophys. Res.*, *110*, D14107, doi: 10.1029/(2004)JD005639.
55. Johnsen, S. J., et al., (2001). Oxygen isotope and paleotemperature records from six Greenland ice-core stations: Camp century, dye-3, GRIP, GISP2, Renland and NorthGRIP. *J. Quat. Sci.*, *16*, 299–307.
56. Joos, F., et al., (2004). Transient simulations of Holocene atmospheric carbon dioxide and terrestrial carbon since the last glacial maximum. *Global Biogeochem. Cycles*, *18*, 1–30. doi: 10.1029/(2003)GB002156.
57. Kageyama, M., et al., (2006). Last glacial maximum temperatures over the North Atlantic, Europe, and western Siberia: A comparison between PMIP models, MARGO sea-surface temperatures and pollen-base reconstructions. *Quat. Sci. Rev.*, *25*, 2082–2102.
58. Kaplan, J. O., et al., (2003). Climate change and Arctic ecosystems: 2. Modeling, paleodata-model comparisons, and future projections. *J. Geophys. Res.*, *108*, 1–27. doi: 10.1029/(2002)JD002559.
59. Kaplan, J. O., Prentice, I. C., Knorr, W., & Valdes, P. J., (2002). Modeling the dynamics of terrestrial carbon storage since the last glacial maximum. *Geophys. Res. Lett.*, *29*, 1–29. doi: 10.1029/(2002)GL015230.
60. Keeling, C. D., & Whorf, T. P., (2005). Atmospheric CO_2 records from sites in the SiO air sampling network. In: *Trends: A Compendium of Data on Global Change* (Vol. 1, pp. 1–23). Carbon dioxide information analysis center, Oak Ridge National Laboratory, U.S. Department of Energy, Oak Ridge, TN.
61. Kennett, J. P., & Stott, L. D., (1991). Abrupt deep-sea warming, paleoceanographic changes and benthic extinctions at the end of the Paleocene. *Nature*, *353*, 225–229.

62. Koch, P. L., Zachos, J. C., & Gingerich, P. D., (1992). Correlation between isotope records in marine and continental carbon reservoirs near the Paleocene Eocene boundary. *Nature*, *358*(6384), 319–322.

63. Koerner, R. M., & Fisher, D. A., (2002). Ice-core evidence for widespread Arctic glacier retreat in the Last Interglacial and the early Holocene. *Ann. Glaciol.*, *35*, 19–24.

64. Koerner, R. M., (1989). Ice core evidence for extensive melting of the Greenland Ice-sheet in the last interglacial. *Science*, *244*(4907), 964–968.

65. Kohfeld, K., & Harrison, S. P., (2001). DIRTMAP: The geological record of dust. *Earth Sci. Rev.*, *54*, 81–114.

66. Kucera, M., et al., (2005). Multiproxy approach for the reconstruction of the glacial ocean surface (MARGO). *Quat. Sci. Rev.*, *24*, 813–819.

67. Kurtz, A. C., Kump, L. R., Arthur, M. A., Zachos, J. C., & Paytan, A., (2003). Early Cenozoic decoupling of the global carbon and sulfur cycles. *Paleoceanography*, *18*(4), 1–33.

68. Lambeck, K., & Chappell, J., (2001). Sea level change through the last glacial cycle. *Science*, *292*(5517), 679–686.

69. Lambeck, K., (2002). Sea-level change from mid-Holocene to recent time: An Australian example with global implications. In: Mitrovica, J. X., & Vermeersen, L. A., (eds.), *Ice Sheets, Sea Level, and the Dynamic Earth* (Vol. 29, pp. 33–50). Geodynamic Series, American Geophysical Union, Washington, DC.

70. Lear, C. H., Rosenthal, Y., Coxall, H. K., & Wilson, P. A., (2004). Late Eocene to early Miocene ice sheet dynamics and the global carbon cycle. *Paleoceanography*, *19*(4), PA4015, doi: 10.1029/(2004)PA001039.

71. Lhomme, N., Clarke, G. K. C., & Marshall, S. J., (2005). Tracer transport in the Greenland Ice Sheet: constraints on ice cores and glacial history. *Quat. Sci. Rev.*, *24*, 173–194.

72. Lisiecki, L. E., & Raymo, M. E., (2005). A Pliocene-Pleistocene stack of 57 globally distributed benthic d18O records. *Paleoceanography*, *20*, PA1003, doi: 10.1029/(2004)PA001071.

73. Liu, Z., et al., (2002). Tropical cooling at the last glacial maximum and extratropical ocean ventilation. *Geophys. Res. Lett.*, *29*, 1409, doi: 10.1029/(2001)GL013938.

74. MacFarling Meure, C., et al., (2006). The law dome CO_2, CH_4, and N_2O ice core records extended to (2000) years BP. *Geophys. Res. Lett.*, *33*, L14810, doi: 10.1029/(2006)GL026152.

75. Machida, T., et al., (1995). Increase in the atmospheric nitrous oxide concentration during the last 250 years. *Geophys. Res. Lett.*, *22*, 2921–2924.

76. Mackay, A., Battarbee, R., Birks, J., & Oldfield, F. E., (2003). *Global Change in the Holocene* (Vol. 1, p. 480). Hodder Arnold, London.

77. Mahowald, N., et al., (1999). Dust sources and deposition during the last glacial maximum and current climate: A comparison of model results with paleodata from ice cores and marine sediments. *J. Geophys. Res.*, *104*, 15859–15916.

78. Mangerud, J., Astakhov, V., & Svendsen, J. I., (2002). The extent of the Barents-Kara ice sheet during the last glacial maximum. *Quat. Sci. Rev.*, *21*, 111–119.

79. Masson-Delmotte, V., et al., (2006). Past and future polar amplification of climate change: Climate model intercomparisons and ice-core constraints. *Clim. Dyn.*, *26*(5), 513–529.

80. McElwain, J. C., & Chaloner, W. G., (1995). Stomatal density and index of fossil plants track atmospheric carbon dioxide in the Paleozoic. *Ann. Bot. (London)*, *76*, 389–395.

81. McManus, J. F., et al., (2002). Thermohaline circulation and prolonged interglacial warmth in the North Atlantic. *Quat. Res.*, *58*, 17–21.

82. Monnin, E., et al., (2001). Atmospheric CO_2 concentrations over the last glacial termination. *Science*, *291*(5501), 112–114.

83. Monnin, E., et al., (2004). Evidence for substantial accumulation rate variability in Antarctica during the Holocene, through synchronization of CO_2 in the Taylor Dome, Dome C and DML ice cores. *Earth Planet. Sci. Lett.*, *224*(1–2), 45–54.

84. Moran, K., et al., (2006). The Cenozoic paleoenvironment of the Arctic Ocean. *Nature*, *441*, 601–605.

85. Mudelsee, M., (2001). The phase relations among atmospheric CO_2 content, temperature, and global ice volume over the past 420 ka. *Quat. Sci. Rev.*, *20*, 583–589.

86. Muhs, D. R., Simmons, K. R., & Steinke, B., (2002). Timing and warmth of the last interglacial period: New U-series evidence from Hawaii and Bermuda and a new fossil compilation for North America. *Quat. Sci. Rev.*, *21*, 1355–1383.

87. Myhre, G., Highwood, E. J., Shine, K. P., & Stordal, F., (1998). New estimates of radiative forcing due to well mixed greenhouse gases. *Geophys. Res. Lett.*, *25*, 2715–1718.

88. NGRIP (North Greenland Ice Core Project), (2004). High-resolution record of Northern Hemisphere climate extending into the last interglacial period. *Nature*, *431*, 147–151.

89. NRC (National Research Council), (2006). *Surface Temperature Reconstructions for the Last 2,000 Years* (p. 196). National Academies Press, Washington, DC.

90. Otto-Bliesner, B. L., et al., (2006b). Last glacial maximum and Holocene climate in CCSM3. *J. Clim.*, *19*, 2567–2583.

91. Otto-Bliesner, B. L., Marshall, S. J., Overpeck, J. T., & Miller, G., (2006a). Simulating Arctic climate warmth and icefield retreat in the last interglaciation. *Science*, *311*, 1751–1753.

92. Overpeck, J., & Trenberth, K. E., (2004). *A Multi-Millennia Perspective on Drought and Implications for the Future* (Vol. 1, pp. 20–30). Proceedings of a joint CLIVAR/PAGES/IPCC workshop, Tucson, AZ. University Corporation for Atmospheric Research, Boulder CO.

93. Pagani, M., et al., (2005). Marked decline in atmospheric carbon dioxide concentrations during the Paleogene. *Science*, *309*(5734), 600–603.

94. Pearson, P. N., & Palmer, M. R., (2000). Atmospheric carbon dioxide concentrations over the past 60 million years. *Nature*, *406*, 695–699.

95. Peltier, W. R., & Fairbanks, R. G., (2006). Global glacial ice volume and last glacial maximum duration from an extended Barbados sea level record. *Quat. Sci. Rev.*, *25*, 3322-3337.

96. Peltier, W. R., (2001). Global glacial isostatic adjustment and modern instrumental records of relative sea level history. In: *Sea Level Rise: History and Consequences* (Vol. 1, pp. 65–95). Academic Press, San Diego, CA.

97. Peltier, W. R., (2002). On eustatic sea level history: Last glacial maximum to Holocene. *Quat. Sci. Rev.*, *21*(1–3), 377–396.

98. Peltier, W. R., (2004). Global glacial isostasy and the surface of the ice-age Earth: The ICE-5G(VM2) model and GRACE. *Annu. Rev. Earth Planet. Sci.*, *32*, 111–149.

99. Peltier, W. R., Shennan, I., Drummond, R., & Horton, B., (2002). On the postglacial isostatic adjustment of the British Isles and the shallow viscoelastic structure of the earth. *Geophys. J. Int.*, *148*(3), 443–475.

100. Pépin, L., Raynaud, D., Barnola, J.-M., & Loutre, M. F., (2001). Hemispheric roles of climate forcing during glacial-interglacial transitions, as deduced from the Vostok record and LLN-2D model experiments. *J. Geophys. Res.*, *106*(D23), 31885–31892.
101. Petit, J. R., et al., (1999). Climate and atmospheric history of the past 420,000 years from the Vostok ice core, Antarctica. *Nature*, *399*, 429–436.
102. Prentice, I. C., & Jolly, D., (2000). BIOME 6000 participants. Mid-Holocene and glacial-maximum vegetation geography of the northern continents and Africa. *J. Biogeogr.*, *27*, 507–519.
103. Rahmstorf, S., (2001). Abrupt climate change. In: *Encyclopedia of Ocean Sciences* (Vol. 1, pp. 1–6). Academic Press, London.
104. Raymo, M. E., & Rau, G. H., (1992). Plio-Pleistocene atmospheric CO_2 levels inferred from POM d13C at DSDP Site 607. *Eos*, *73*, 95.
105. Raymo, M. E., Grant, B., Horowitz, M., & Rau, G. H., (1996). Mid-Pliocene warmth: Stronger greenhouse and stronger conveyor. *Mar. Micropaleontol.*, *27*(1–4), 313–326.
106. Raynaud, D., Chappellaz, J., Ritz, C., & Martinerie, P., (1997). Air content along the Greenland Ice Core Project core: A record of surface climatic parameters and elevation in central Greenland. *J. Geophys. Res.*, *102*, 26607–26614.
107. Rind, D., & Chandler, M. A., (1991). Increased ocean heat transports and warmer climate. *J. Geophys. Res.*, *96*, 7437–7461.
108. Rosell-Mele, A., et al., (2004). Sea surface temperature anomalies in the oceans at the LGM estimated from the alkenone-UK'37 index: comparison with GCMs. *Geophys. Res. Lett.*, *31*, L03208, doi: 10.1029/(2003)GL018151.
109. Rostami, K., Peltier, W. R., & Mangini, A., (2000). Quaternary marine terraces, sea-level changes and uplift history of Patagonia, Argentina. Comparisons with predictions of ICE-4G (VM2) model of the global process of glacial isostatic adjustment. *Quat. Sci. Rev.*, *19*, 1495–1525.
110. Royer, D. L., (2006). CO_2-forced climate thresholds during the Phanerozoic. *Geochim. Cosmochim. Acta*, *70*(23), 5665–5675.
111. Royer, D. L., et al., (2001). Paleobotanical evidence for near present-day levels of atmospheric CO_2 during part of the tertiary. *Science*, *292*(5525), 2310–2313.

112. Royer, D., (2003). Estimating latest cretaceous and tertiary atmospheric CO_2 from stomatal indices. In: Wing, S. L., & Gingerich, P. D., (eds.), *Causes and Consequences of Globally Warm Climates in the Early Paleogene.*

113. Ruddiman, W. F., & Thomson, J. S., (2001). The case for human causes of increased atmospheric CH_4. *Quat. Sci. Rev.*, *20*(18), 1769–1777.

114. Ruddiman, W. F., (1997). *Tectonic Uplift and Climate Change* (Vol. 1, pp. 500–535). Plenum Press, New York.

115. Schwander, J., et al., (1993). The age of the air in the firn and the ice at Summit, Greenland. *J. Geophys. Res.*, *98*(D2), 2831–2838.

116. Shackleton, N. J., (1977). Carbon-13 in Uvigerina: Tropical rainforest history and the equatorial Pacific carbonate dissolution cycles. In: Andersen, N., & Malahoff, A., (eds.), *The Fate of Fossil Fuel CO_2 in the Ocean* (Vol. 1, pp. 401–428). Plenum, New York.

117. Shackleton, N. J., (2000). The 100,000-year ice-age cycle identified and found to lag temperature, carbon dioxide, and orbital eccentricity. *Science*, *289*, 1897–1902.

118. Shackleton, N. J., Hall, J. C., & Pate, D., (1995). Pliocene stable isotope stratigraphy of ODP Site 846. In: *Proceedings of the Ocean Drilling Program, Scientific Results* (Vol. 138, pp. 337–356). Ocean Drilling Program, College Station, TX.

119. Shin, S. I., et al., (2003). A simulation of the last glacial maximum climate using the NCAR CSM. *Clim. Dyn.*, *20*, 127–151.

120. Siddall, M., et al., (2003). Sea-level fluctuations during the last glacial cycle. *Nature*, *423*, 853–858.

121. Siegenthaler, U., et al., (2005a). Stable carbon cycle-climate relationship during the late Pleistocene. *Science*, *310*(5752), 1313–1317.

122. Siegenthaler, U., et al., (2005b). Supporting evidence from the EPICA Dronning Maud Land ice core for atmospheric CO_2 changes during the past millennium. *Tellus*, *57B*(1), 51–57.

123. Sloan, L. C., Crowley, T. J., & Pollard, D., (1996). Modeling of middle Pliocene climate with the NCAR GENESIS general circulation model. *Mar. Micropaleontol.*, *27*, 51–61.

124. Sluijs, A., et al., (2006). Subtropical Arctic Ocean temperatures during the Paleocene/Eocene thermal maximum. *Nature*, *441*, 610–613.

125. Spahni, R., et al., (2003). The attenuation of fast atmospheric CH_4 variations recorded in polar ice cores. *Geophys. Res. Lett.*, *30*(11), 1–30. doi: 10.1029/(2003)GL017093.

126. Spahni, R., et al., (2005). Atmospheric methane and nitrous oxide of the late Pleistocene from Antarctic ice cores. *Science, 310*(5752), 1317–1321.

127. Stauffer, B., Fischer, G., Neftel, A., & Oeschger, H., (1985). Increase of atmospheric methane recorded in Antarctic ice core. *Science, 229*(4720), 1386–1388.

128. Steele, L. P., et al., (1992). Slowing down of the accumulation of atmospheric methane during the 1980s. *Nature, 358*, 313–316.

129. Stocker, T. F., & Monnin, E., (2003). Past rates of carbon dioxide changes and their relevance for future climate. *Pages News, 11*(1), 6–8.

130. Svensen, H., et al., (2004). Release of methane from a volcanic basin as a mechanism for initial Eocene global warming. *Nature, 429*, 542–545.

131. Tans, P. P., & Conway, T. J., (2005). Monthly atmospheric CO_2 mixing ratios from the NOAA CMDL carbon cycle cooperative global air sampling network, 1968–2002. In: *Trends: A Compendium of Data on Global Change* (Vol. 1, pp. 1–24). Carbon dioxide information analysis center, Oak Ridge National Laboratory, U. S. Department of Energy, Oak Ridge, TN.

132. Tarasov, L., & Peltier, W. R., (2003). Greenland glacial history, borehole constraints, and Eemian extent. *J. Geophys. Res., 108*, 2143, doi: 10.1029/(2001)JB001731.

133. Taylor, K. E., et al., (2000). Analysis of forcing, response, and feedbacks in a paleoclimate modeling experiment. In: Braconnot, P., (ed.), *Proceedings of the Third Paleoclimate Modelling Intercomparison Project (PMIP) Workshop, La Huardière, Canada* (Vol. 1007, pp. 43–49). WCRP-111, WMO/TD-No. 1007, World Meteorological Organization, Geneva.

134. Thomas, E., (2003). Extinction and food at the sea floor: A high-resolution benthic foraminiferal record across the initial Eocene thermal maximum, Southern ocean site 690. In: Wing, S., Gingerich, P., Schmitz, B., & Thomas, E., (eds.), *Causes and Consequences of Globally Warm Climates of the Paleogene* (Vol. 369, pp. 319–332). Special Paper, Geological Society of America, Boulder, CO.

135. Thompson, R. S., & Fleming, R. F., (1996). Middle Pliocene vegetation: Reconstructions, paleoclimate inferences, and boundary conditions for climate modeling. *Mar. Micropaleontol., 27*, 27–49.

136. Thompson, R. S., (1991). Pliocene environments and climates in the Western United States. *Quat. Sci. Rev.*, *10*, 115–132.
137. Toracinta, E. R., Oglesby, R. J., & Bromwich, D. H., (2004). Atmospheric response to modified CLIMAP ocean boundary conditions during the last glacial maximum. *J. Clim.*, *17*, 504–522.
138. Tripati, A. K., & Elderfield, H., (2004). Abrupt hydrographic changes in the equatorial Pacific and subtropical Atlantic from foraminiferal Mg/Ca indicate greenhouse origin for the thermal maximum at the Paleocene-Eocene Boundary. *Geochem. Geophys. Geosystems*, *5*, 1–20. doi: 10.1029/(2003)GC000631.
139. Waelbroeck, C., et al., (2002). Sea-level and deep-water temperature changes derived from benthic foraminifera isotopic records. *Quat. Sci. Rev.*, *21*(1–3), 295–305.
140. Wing, S. L., et al., (2005). Transient floral change and rapid global warming at the Paleocene-Eocene boundary. *Science*, *310*(5750), 993–996.
141. Wyputta, U., & McAvaney, B. J., (2001). Influence of vegetation changes during the last glacial maximum using the BMRC atmospheric general circulation model. *Clim. Dyn.*, *17*, 923–932.
142. Yapp, C. J., & Poths, H., (1992). Ancient atmospheric CO_2 pressures inferred from natural goethites. *Nature*, *355*, 342–344.
143. Zachos, J. C., et al., (2003). A transient rise in tropical sea surface temperature during the Paleocene-Eocene thermal maximum. *Science*, *302*(5650), 1551–1554.
144. Zachos, J. C., et al., (2004). *Early Cenozoic Extreme Climates: The Walvis Ridge Transect, Sites 1262-1267* (Vol. 208, pp. 1–24). Proceedings of the ocean drilling program, initial reports, ocean drilling program, college station, TX.
145. Zachos, J. C., et al., (2005). Rapid acidification of the ocean during the Paleocene-Eocene thermal maximum. *Science*, *308*(5728), 1611–1615.
146. Zachos, J., et al., (2001). Trends, rhythms, and aberrations in global climate 65 Ma to present. *Science*, *292*(5517), 686–693.
147. Zhang, D. E., (2005). Severe drought events as revealed in the climate record of China and their temperature situations over the last 1000 years. *Acta Meteorol. Sin.*, *19*(4), 485–491.

Chapter 2

History of Paleoclimatology

CONTENTS

2.1. Introduction 46
2.2. Beginning of Reason 47
2.3. The Scientific Strategy 47
2.4. Concept of Climatic Cycles 49
2.5. Commercialization of Paleoclimatology 53
2.6. Actualism and Uniformitarianism 54
2.7. The Theory of Glacial 55
2.8. From Accumulated Ice to Glaciation 56
2.9. Pre-Quaternary Eras of ICE 57
2.10. Eventually: Climatic Variations 59
2.11. Cosmic Variables 60
2.12. Elements in the Galactic Arms 61
2.13. Solar Radiation 61
2.14. Orbital Dissimilarities 62
2.15. Events of Asteroid Collision 63
2.16. Endogenic Global Heating 65
2.17. Plate Tectonic and Relocation of Oceans and Continents 66
2.18. Orographic Reaction to Eustasy and Plate Tectonics 68
2.19. Uplift of Mountain and Global Cooling 69
2.20. Oceanic Circulation 70
2.21. Sea Ice and the Glacier Variation 71
2.22. Variations in Atmospheric Composition 72
2.23. The Moon, Lunar-Solar Tides, and Associated Solar, Terrestrial, and Planetary Cycles 74
2.24. Volcanic Eruptions and the Desert Dust 77
2.25. Paleosols and Loess 78
2.26. Eolianite and Eolization 79
2.27. Levels of the Lake and Fluvial Signs 80
References 83

2.1. INTRODUCTION

The chapter under focus will treat in sequence (a) the discovery stage of the proof and (b) the experimental and causative stage. The elemental idea of climate variation possibly progressed in the form of documentaries in ancient Egypt, Mesopotamia, China, and the Indus Valley, where unusual floods or prolonged famines were suffered. Interannual to a long change in the amounts of Nile floods were less or more resembled by the lucks of consecutive dynasties (Fairbridge, 1967; Thunell et al., 1977). The famine and scarcity times were graver socially as compared to the episodic occurrences of too many floods. Initial sections of the Judeo-Christian Bible, particularly Genesis, described the story of 40-day rain, occasioning in the floods of Noah and the consequent retrieval. Some of the other eastern cultures comprised comparable customary tales. It can thus be the rational assumption that the idea or concept of periodic climate variation was not only definitely developed in the folktale of humans, but explained by periodic fortification offered by inconsistent climatic happenings (Ryan and Pitman, 1999).

Throughout the past 10,000 years, circumstances of climate have disturbed the history of humans on several scales, proof for which is exposed with growing incidence. The part of external occasions, small or large, as contributing aspects in the history of humans, therefore became the dominant characteristic in Aristotelian philosophy, and the strategy inclined to influence history for the coming 2 eras (Schramm, 1963). For instance, the expansion of gigantic pro-glacial waters across North America throughout the withdrawal of the Laurentian Ice Sheet was escorted by the isostatic crustal reverberation that ultimately slanted these huge lakes towards the sea, occasioning in freshwater floods that blowout and ultimately watered-down neighboring surfaces of the ocean, instigating paleoclimatic catastrophes with worldwide consequences. This might have added to the widespread tradition of a flood (Genty and Quinif, 1996; Bradley, 1999).

One more single-event occurrence was the sinking of Euxinic waters in the region of the Black Sea when the rise of the post-glacial level of sea exceeded the maximum level of Bosporus (Fairbridge, 1966). A current oceanographic investigation has disclosed ample proof of the past seaside, the human inhabitants, and its wildlife (Ryan and Pitman, 1999; Bralower, 2002).

From the ancient viewpoint, the denial of disastrous hypotheses demonstrates the distinctive traditionalism of geologists elevated on the Lyellian uniformitarianism thinking. The ultimate recognition of such outrageous hypotheses has slowly led to the altered thinking of neocatastrophism (Cerling, 1991; Caillon et al., 2003).

2.2. BEGINNING OF REASON

Even before the Lyell time (1797 to 1875), slow promotion of the anti-miracle and the anti-catastrophic group had started. By the 18th century, interpretations might frequently be supplemented by geological fieldwork and chemistry, and the latter offered the never-ending inconsistencies of paleoclimatology. At the initial stages, the viewers were similar to the stalker of anthropology. These observers just discovered. They began documenting their interpretations on maps. Particular ideas such as petrology, structure, stratigraphy, paleontology, and mineralogy developed in the 19th century (Chandler et al., 1994; Chappellaz et al., 1997). Viewers of the 18th to initial 19th centuries were generally divided into 2 categories: (a) a gentlemen traveler, *tha*t is, notable with private aids who discovered the flora and fauna due to interest; and (b) the people supported commercially who were goal-oriented, examining explicitly for resources like coal or the metallic reserves like gold, iron, silver, lead, and zinc; or the qualified, who were designing water-supplies or canals. The viewers who inquired queries regarding the circumstances of climate specified by that or this substitution at the formation time had literacy and leisure. An initial conversation of the scientific queries was made by Emanuel Swedenborg, the Swedish theologian and scientist, who accredited raised seashores to advanced aridity of the globe ocean after the floods of Biblical (Dowsett and Cronin, 1990; Dlugokencky et al., 1994).

Strangely, the individuals requesting these queries were often participants of the clerics, who were not just knowledgeable but whose responsibilities demanded substantial traveling. Therefore, Nicolaus Steno, the founder of stratigraphy, even though qualified in flat topography of Denmark, consumed most of his profession in northern Italy, where Nicolaus discovered time to precisely perceiving and document the structures and stratification of rock creations (Dowsett et al., 1996, 2005).

2.3. THE SCIENTIFIC STRATEGY

Schwarzbach, a historian of earth science acknowledged an English physicist, Robert Hooke (1635 to 1703), as the leading researcher in paleoclimatology. He perceived huge ammonites and enormous turtles in Portlandian of Dorset in England and established that they specified the warm climate once. This he assumed might be explained only by the change in the axis of Earth (Enting, 1987; EPICA Community Members, 2004).

One more creative discoverer was George L. Leclerc (1707 to 1788), a distinguished French researcher. In the book "Epoques de la Nature" (1778), he defined the slow cooling of the Earth when molten, the cooling that started at poles. Previously warmer temperatures at the elevated latitudes were established by findings of the huge mammal remainders, like rhinoceroses and elephants, in the areas; such kinds of living beings are discovered in warm climatic conditions nowadays (Fairbanks, 1989; Etheridge et al., 1996).

Another individual was Thomas Jefferson (1743 to 1826), 3rd president of U.S., who had worked as the ambassador to France and also became accustomed to Buffon. In the book "Notes on the State of Virginia" (1785), Thomas conferred the finding of huge bones in what he considered as anomalously elevated latitudes. Thomas assumed this might probably be described by the variation in the angle of the ecliptic of Earth, and therefore the spin axis slant that governs the yearly distribution of seasons. Presently, Southern Hemisphere summer corresponds with the closest approach of Earth to the star Sun and summer of the Northern Hemisphere arrives when the Earth and Sun are farthermost apart from each other. Nearly 9,000 years ago, summer of the northern came when the orbit of Earth was nearby the star sun, thus the mid-latitude means temperatures of summer were 2°C to 3°C warmer as compared to the present. The question of North America's mammoths could instead be described by the variation in the preferred habitat of creatures (Fischer and Wefer, 1999; Ferretti, 2005).

Alexander von Humboldt (1769 to 1859) was the typical gentleman-traveler-scientist, a well-known German researcher of the 19th century, particularly for the efforts in South America. Alexander resided in Paris for almost 20 years and therefore had the ideal luck to attain an approach to the scientific libraries and also to modern thinkers and naturalists like Buffon. In order to start with, in the year 1799, though, Alexander submitted to A. G. Werner's Neptunian idea of water laid basalts and granites, an incorrect concept, albeit donating to the warmer Earth. Far ahead, after perceiving dynamic volcanoes, albeit was perhaps the 1st to identify the probable impact of dust and ash clouds on the climate cooling (1823).

2.4. CONCEPT OF CLIMATIC CYCLES

The typical idea of climate cycles is perhaps that accredited to Joseph, an associate of the Israelite colony in Egypt, possibly around the 14th century BC. Joseph warned of seven years of the bounty trailed by seven years of poor circumstances. The message was very obvious: climates vary, so, be careful and plan. The Egyptological record comprise proof of high-water symbols on the columns of the temple, interchanging with symbols of erosional segmentation in the terraces of the Nile, which normally signify to eras when the water flow was slow. Papyrus explanations can be validated now by geological investigations (Fairbridge, 1962; Flückiger et al., 1999, 2002).

Organized computation of the cycles of Nile flood was started during the period of Moslem, with numerous centuries of precise information. Long and short cycles have generally been recognized (Hassan, 1981), which contemplate rainfall of monsoon, primarily in Ethiopian waters of the Blue-Nile. The spectrum examination (Currie, 1995) reveals lunar (18.60 years) and solar (11.0 years) periodicities (Figure 2.1).

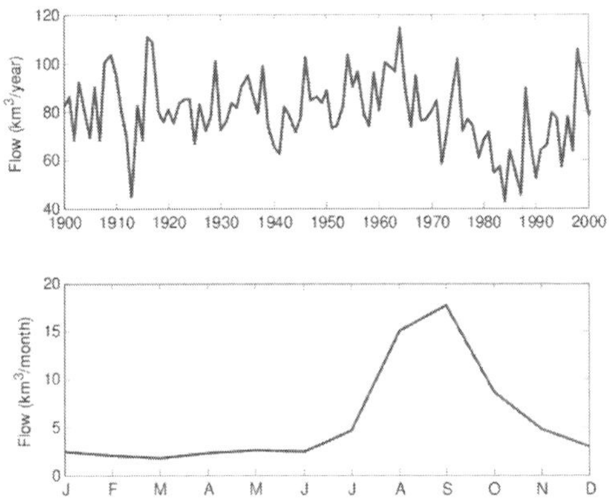

Figure 2.1. Cycles of the flow of the Nile during 10 years (on top) and the complete year (on bottom).

The first conversation of higher shorelines seems to have been made by Emanuel Swedenborg (1688 to 1772), as stated, who accredited them to signs of previous levels gathered or worn throughout the advanced aridness of the ocean of the world after the Biblical flood. The experimental strategy was developed by Carl Linneus (1707 to 1778), the Swedish professor, cooperatively with the physics professor Anders Celsius. Linneus was quite famous for his organized biologic classification. Together they gave an extraordinary contribution to the history of climate when they invalidated the theory of Desiccation, which generally had become common in the 18th century. This philosophy was given by Bishop of Abo and centered on the slow appearance of the Scandinavia coasts, which Bishop and his colleague theologians inferred as convincing proof for the tale of the flood of Noachian and the consequent collapse. Celsius and Linneus invented the practical experiment in order to validate the theory of Biblical Desiccation. After hearing that southern Sweden was evolving more gradually as compared to the country's northern region, they determined on the summer holidays and molded the series of marks of the level of water along the shore. Every site was visited again yearly for 10 years and the results were announced then. In the north near the Bothnia Gulf, the shore seemed to be expanding at 6 to 8 millimeters annually, whereas in the south, the coast of the Baltic was falling gradually; in some regions it was negative. The desiccation of the world's ocean would be identical; however, the comparative motion should contemplate the movement of the crust of Earth. This practical experiment was perhaps the 1st experiment in paleoclimatology and crustal variation (Francois et al., 1998, 1999).

Anders Celsius (1701 to 1744), a Swedish astronomer and physicist at Uppsala simultaneously as the Linneus, was renowned for his interpretations of aurora borealis (1716 to 1732). In addition to assisting to compute the meridian arc in Lapland, he became famous for evolving the scale of centigrade, later called the Celsius scale for thermometers. Precisely attuned instruments were given to the divisions of physics of all main universities. Temperatures could be quantitatively measured for the first time (Freeman and Hayes, 1992; Guo et al., 2004).

In association with the experiments of Linneus-Celsius must be stated the Baron Christian Leopold von Buch experience (1774 to 1853), a well-known geologist (German). He made numerous boat tours around northern Norway, along with in the Baltic. He validated that the coast was expanding at Åbo. Northern Norway is subjected to remarkable coast erosion and is unprotected from drastic rainstorm waves of North Atlantic. The uplift of crust in northern Norway, though, created the succession of appeared seats in the shape of the staircase. These melodramatic terranes are recurrent along the coastline for almost more than 1,000 km. In the year 1810, Von Buch accredited this uplifting to the pulsatory magmatic reasons, like he had observed functioning in the Islands of Canary and some other volcanic areas (Joos et al., 2004; Kageyama et al., 2006).

Reusch, in the year 1894, invented the term strand flat, for the broad platforms of the shore. Though, storm waves were not responsible as shore corrosion is discovered on the internal edges of fjords. In order to describe them, Nansen (1922) suggested the 2-stage procedure: (a) Ice action above or at the level of sea, trailed by (b) freezing and transportation of remains by the moving ice. First interpretations by Lyell in Quebec assisted his drift philosophy. Traces of the fossil strand flats and boulders moved by ice are observed on the shores of France and England (Fairbridge, 1977; Kaplan et al., 2002, 2003). Could cyclical procedures be initiated extra terrestrially? The concept that the star Sun may play a part to trigger the climatic happenings was generally to step on the subtle ground. When Galileo Galilei (1564 to 1642) 1st perceived sunspots, it disguised to few theologians a deficiency in the creation of God, the star Sun, and normally that was an abomination. Astronomy steered nonetheless to the concept of foreseeable cycles. When Edmond Halley (1656 to 1742) effectively predicted the return of the eponymous comet in the year 1758, the situation was developed. The comet reverted in 1986, 1910, and 1835. He placed the basis for the novel geomagnetism, which ultimately facilitated to relate of endogenetic powers with paleo-climatology. William Herschel (1738 to 1822), an astronomer, suggested that not just did the sunspot vary cyclically, but they might also link to some climatic variations that could disturb the wheat price (Kennett and Stott, 1991; Keeling and Whorf, 2005).

Systematic interpretations of sunspots were initiated in the year 1826 by S. H. Schwabe, who declared an approximately 10-year cycle in the year 1843. Edward Sabine at that moment had generally been observing geomagnetic variations in Canada and discovered an analogous fluctuation, even inferring at the mutual cause. This was recommended autonomously by A. Gautier and R. Wolf, in the year 1852, in France and Switzerland, correspondingly, and also by R. C. Carrington in England, in the year 1860.

The century of disagreement regarding Sun-climate associations has trailed. Ample literature gathered over the eras, briefed in cycles of the sunspot (Schove, 1983). Presently, the idea of the solar compelling cycle is starting to be acknowledged, however, the "why" and "how" are still indefinable (Bond et al., 2001; Rind, 2002; Shindell et al., 2001). Systematic temperature measurements unlocked the way to the concept of climate cycles. The applied phase was made feasible by Celsius in the 18th century, and 2 centuries after came to the 1st examination of the temperature and time sequence that visibly demonstrated climate cycles (Figure 2.2) (Fairbridge, 1968; Koch et al., 1992).

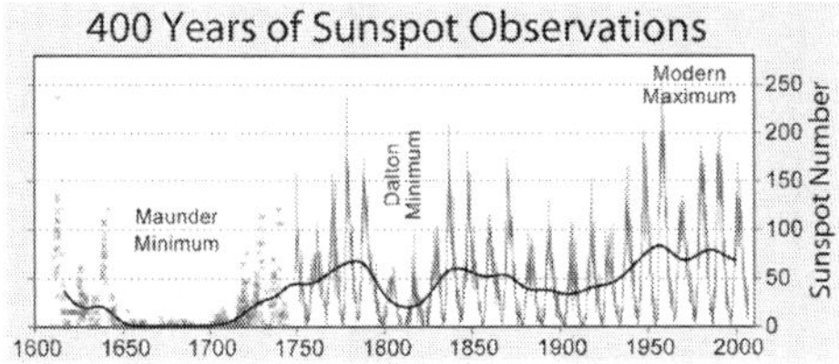

Figure 2.2. Sunspot observation data for the last four years.

Source: http://ossfoundation.us/projects/environment/global-warming/natural-cycle.

The time series of Labrijn disclosed cycles of temperature conforming to 22-year solar radiation fluctuations and magnetic reversals. These interpretations were applied to the area of North Atlantic-Europe covered by the recordings of temperature started by Anderus Celsius, however, they made the path to 20th-century thinking (Koerner, 1989; Mahowald et al., 1999).

The presence of the 22-year cycle was confirmed by Schuurmans (Fantechi and Flohn, 1984). This cycle was strengthened when the small faults triggered by urban heating consequences were eliminated competently in the 250-year study of the Stockholm temperatures carried out by Anders Moberg. The statistical process that made a study of Labrijn so extraordinary utilized months of summer (June to August) minus the months of winter (December to February) in order to make the continentality record (Mangerud et al., 2002; Masson-Delmotte et al., 2006).

One hint to solar and climate associations is available in the star Moon. The part of the star Moon in the cycles of weather has generally been developed in traditional stories, for worse or better. George Schubler. a physics professor at the University of Tubingen, presented the paper in the year 1836 comprising statistical examinations of his 60 long years of methodical computations of the rainfall, which revealed 18.60 years cycle of lunar nodal (Fairbridge and Shirley, 1997). The worth stating is a 15th-century agronomy paper that described the benefit in the time of seeding due to the rainfall hints of the 18.7 years cycle. Arthur Schuster of the United Kingdom, famous for the discovery of time-series periodogram, also printed on solar and the lunar periodicities in earthquakes and also on the impact of planets on the creation of sunspots. In the 20th century, lunar cycles were revealed by R. G. Currie in almost all climatic and geophysical time-series (Currie, 1995; McElwain and Chaloner, 1995).

Revolutionary discoveries of William Gilbert (1544 to 1603) in geomagnetism ultimately led to the comprehension of aurora australis and the aurora borealis as terms of the magnetic field of Earth in the expression of a primarily enigmatic solar force. The comprehensive analysis of this subject's history was carried out by Siscoe in the year 1980, developing their associations with the cycles of sunspot, but Siscoe set aside particular associations with the geomagnetic field and the solar wind (McManus et al., 2002).

Scientific interpretations of the geomagnetic field and the aurora were started in the 19th century in Scandinavia. Due to the elevated latitude, the mutual tradition of Sweden and Norway had generally been perceiving these spectacles for numerous centuries. The pastors of the state recorded numerous sightings and then were examined by two well-known scientists Svante Arrhenius and Nils Ekholm. The two scientists recognized the 11-year long solar cycle and also the 18.60-year long lunar cycle, along with the climatic relationships (Ekholm and Arrhenius, 1898). Currently, almost a century after, the important paper is usually never quoted and the associations are comprehended very little (Monnin et al., 2001, 2004).

2.5. COMMERCIALIZATION OF PALEOCLIMATOLOGY

Commercialization of the scientific belief became the trend in the initial 19th century in North America and Europe. Hugh Miller completed discoveries of fossils that became the foundation for paleoclimatic and paleogeographic reconstructions in the writings. These writings were demonstrated to be very stimulating for the non-technical addressees (Mudelsee, 2001; Moran et al., 2006).

Oswald Heer (1809 to 1883), a naturalist and botany professor, discovered proof of the hot climate through the interglacial interim in Alps. The book of Oswald Heer "The Primeval World of Switzerland" was very broadly read and aided launch him as amongst the originators of paleogeography and paleoclimatology. By supposing that petrified plants might be utilized as equivalents for contemporary ones, habitats of the current-day could be utilized for calculating alternative rainfall and temperatures. Keeping in thought that throughout the Cenozoic the Tethys Sea route was exposed to the Pacific and Indian Ocean, calculation of Oswald of the mean heating of 7.0°F (3.50°C) at the particular time was perhaps the best calculation (Brooks, 1949). The intervals of hot paleoclimate in a Cenozoic might seem to have generally been accomplished by 3 synergetic overall paleogeographic aspects: (a) open Tethyan sea route globally; (b) restricted orographic blocking; (c) limited Antarctic Mountain glaciers (Figure 2.3) (Fairbridge, 1961; Muhs et al., 2002).

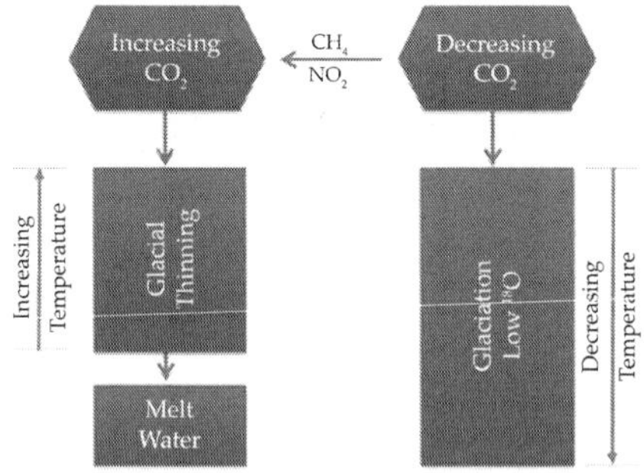

Figure 2.3. Importance of paleoclimatology.

Source: https://link.springer.com/referenceworkentry/10.1007%2F978-1-4614-6431-0_56-1.

2.6. ACTUALISM AND UNIFORMITARIANISM

The Scottish farmer James Hutton (1726 to 1797) examined climate as a worldwide occurrence, even though the work was obscured by the ground-breaking part in geology. Hutton's paper Theory of the Earth (1785) attempted to leave the laws by which Earth history's dynamics could be clarified. In order to describe the erosion of the surface of Earth, he presented the concept of atmosphere and the climate, and a year before Hutton defined the Theory of Rain. The theory comprised the overall climate review, identifying that the humidity of the atmosphere was the main key to precipitation and that the rainfall should happen when the mass of hot air comes across the cooler one. The priceless outline of his concepts was published in 1803 by John Playfair. James Hutton donated primarily to the 18th-century idea that ultimately was known as uniformitarianism. Lyell gave reasons that the principles of nature were everlasting and immutable, which allows one to consider by resemblance. The typical saying is: "the Present is the Key to the Past." While Nicolaus Steno found the concept of time-ordering or sequence, it was James Hutton, after the discovery of the stratigraphic uniqueness, who recognized that geological procedures should not be understood as the continuum however as susceptible to regular revolutions or upsets. Therefore, the idea of the cycle of processes or events appeared, to be associated by Lyell in the stratigraphic series. It was not, though, till the late 20th and early 21st centuries that the sequences became sheltered into the paleoclimatology outline, when Frederick Zeuner (1905 to 1963) and Milutin Milankovitch (1879 to 1958) developed, correspondingly, astronomic compelling possibilities, having the stratigraphic illustration.

A marginally diverse idea of uniformitarianism and the terminology had appeared in eastern Europe and Russia, the term "actualism" being invented by M. V. Lomonosov. As conceived, actualism was an applied approach. It was thus the analytical tool centered on laws originated from measurable representations and would validate extrapolations along with events of the past. On the other hand, uniformitarianism was the hypothesis of geohistory, susceptible to numerous possible hazards of clarification (Albritton, 1975). Presently, the actualistic strategy is favored.

2.7. THE THEORY OF GLACIAL

This concept was believed to become the preoccupation of the scientific world throughout the 19th century. Worthy summaries are normally to be discovered in Schwarzbach (1974); Flint (1971); and Imbrie and Imbrie (1979, 1986).

This philosophy received a substantial shot when the Swiss Natural Science Society in 1817 proclaimed the competition for associates to exhibit the previous addition of Alpine glaciers. The theory was compensated in the year 1822 by the key lecture given by Ignatz Venetz, a Swiss engineer. The subject was considered eagerly by Louis Agassiz, a geologist, who ultimately came to educate at the Harvard University (Figure 2.4).

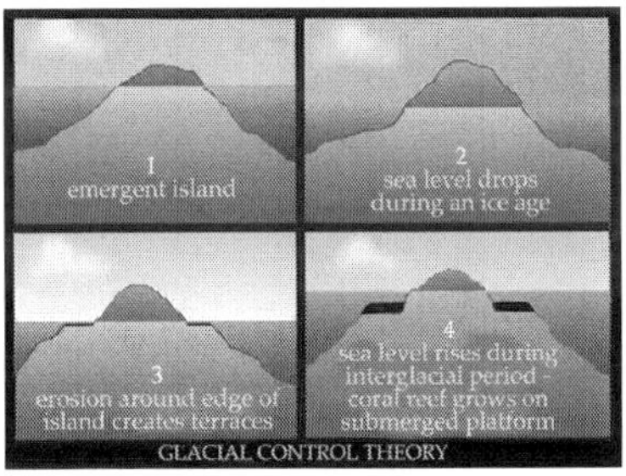

Figure 2.4. Phases of the glacial control theory.

Source: https://www.coraldigest.org/index.php/GlacialActivity.

Scandinavia was the capable resource of an erratic block, but its glaciers were less known and smaller as compared to those of Switzerland. However, Jens Esmark (1762 to 1829) suggested the previous glacial source in 1824 for widespread erratics of northern Europe.

Now in history, establishment hit back, by the colorful and powerful pen invented by Reverend Dean William Buckland (1784 to 1856), the first geology professor at Oxford University. Considering that his income was dependent on the religious adherences, it should not be any shock that the flood of Noachian must be existent in his public speeches. Buckland invented the term diluvium (formation of flood) for accumulations of Pleistocene ice era and the tag came into extensive usage in Europe, recalled into the 20th century. The social nature and eagerness of Buckland brought him criticism; however, he steadfastly developed the idea of stratigraphic paleontology, conflicting categorizations of which establish the main fundamental of paleoclimatology.

2.8. FROM ACCUMULATED ICE TO GLACIATION

The concept of glacial was amongst the extraordinary uprisings in mind that relate to paleoclimatology. The query of uneven rocks to the continental glaciation demands for the great increase in scale, however, it was hardly more than one era before this highbrow leap was accomplished. The basic answer was field illustration. On Alpine expeditions in the year 1836, Jean de Charpentier and Ignatz Venetz convinced Louis Agassiz of previous development of mountain glaciers all over the nearby low-lying topography, and in the year 1837, Agassiz gave his theory of the Great Ice Era. His systematic communication with the geologists caused Alpine expeditions by Buckland and the discussion expeditions by Agassiz to Wales, England, and Scotland. Lyell has displayed the proof, and ultimately harmony prevailed. Nearly the entire British Islands should have been glaciated as stated by T. F. Jamieson. The first examination in Britain of the subject was done by James Geikie in the year 1874, who had been a geology professor. Important hints were also given by North America (Flint, 1971).

Charles Lyell, was not an immediate convert. Lyell made 2 expeditions to North America, comprising 1 trip to St. Laurence valley and the second to the Bay of Fundy in Canada. In these areas, Lyell perceived glacial rocks implanted in moving ice and in 1st version of the book "Principles of

Geology," he accredited all of the uneven rocks to moving ice. Certainly, for a long time, all of the artificial accumulations of obviously glacial sources were categorized in U.S. and Britain as drift. Drift became separated from moving ice in the later versions of Charles Lyell. In the U.S., it assisted as the substitute for glacial accumulations, and the driftless region is yet acknowledged in an upper Midwest (Flint, 1971).

The greeting of glacial theory was not rapid in Germany (Schwarzbach, 1974). However, the German term Eiszeit was an expression utilized in the year 1837 by Karl Schimper, although it was rapidly adopted globally. Geology professor in Berlin, Von Buch, had a huge impact but well-thought-out the uneven rocks to be moved by flow of mud. In the letter of 1850 (Schwarzbach, 1974), Buch specifies to the theory of glacial as "the strange aberration of the human mind." Ultimately the idea became generally acknowledged in Germany after the visit in the year 1875 to Berlin by Otto Torell (1828 to 1900), a Swedish geologist.

Charles MacLaren, in the year 1842 in Scotland, discussed that if the enormous amount of moving ice had generally been consequential of evaporation from the oceans, then the sea level at the maximum of glacial should have been lesser as compared to today. Unluckily he misjudged the volume of former ice and computed the sea-level fall at nearly 801 feet, which demonstrated to be slightly overstated. However, the principle was significant and the worldwide important phenomenon was established now, though it was nearly 100 years before this was integrated into geological concluding. Not just the enormous climatic variations were probable, however, they also disturbed the level of the sea and the comparative size of oceans and continents from which the climatic feedback procedures were unavoidable (Valentine and Moores, 1970).

2.9. PRE-QUATERNARY ERAS OF ICE

Towards the conclusion of the 19th century, additional, widespread glaciations were found to generally have happened formerly to the Quartet Great Ice Era. Permo-Carboniferous Age of Ice was the first to be discovered. Proof for the glaciation was broadly met with astonishment, partially since the circumstances were frequently nearby the equator and some more sites were also included, it was observed that they were crossing the width of the whole hemisphere (Figure 2.5) (Du Toit, 1937).

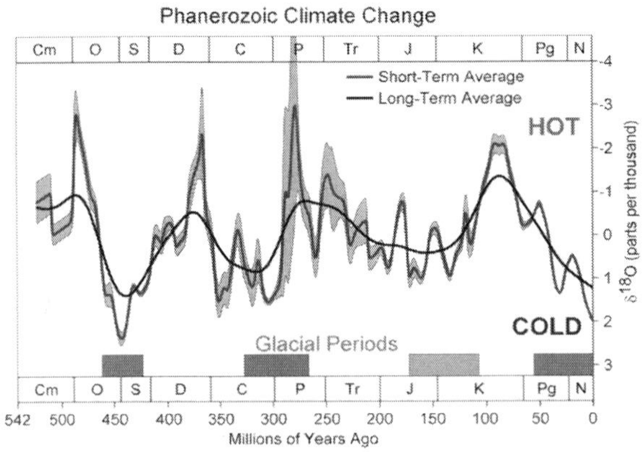

Figure 2.5. Chronology of glaciation.

Source: https://en.wikipedia.org/wiki/Timeline_of_glaciation.

Probably amongst the bravest doings that can be envisaged was when the very young novel entrance to Geological Survey of India, in the year 1856, W. T. Blanford, proclaimed that he had found indisputable proof of the previous age of ice in the Permo-Carboniferous deposits of Deccan, at the distance of just over 2,000 kilometers from equator. At this point, predetermined planetary orbits, predetermined continents, and predetermined concepts were instead established rigidly. Though, cynics might be surprised to know that Blanford later became Manager of the Survey, his astonishing innovation being validated later somewhere else in India, along with within all of the regions of the Southern Hemisphere. This creation of glacial, known as Talchir Tillite, over 100 meters thick in regions, became the main aspect in the idea of the previous massive continent, known as Gondwanaland by Suess (1831 to 1914), fragments of which currently overlap the equator.

Permo-Carboniferous accumulations of glacial and striated paths were found all over the Gondwanaland (Frakes, 1979). These deposits were informed in Australia, Africa, and Saudi Arabia (Crowell, 1999). These deposits were found in South America, and lastly in Antarctica. Factually, these discoveries produced numerous interesting and striking theories. The most important of the theories was the Theory of Continental Drift given by Alfred Wegener which demonstrated to be the precursor of the current plate-tectonic hypothesis (Emiliani, 1992). In this manner, with the help of geophysics and marine geology, an answer was discovered for what was an unsolvable issue in paleoclimatology.

Intervals of the great ice era were consequently found in former geologic eras. One of these intervals was of the Late Ordovician and this interval was ultimately discovered to have spread from Brazil via West Africa and entirely across the Sahara to Saudi Arabia and Ethiopia. Fascinatingly it was found in the 1970s, firstly by the geologists of France. Its existence was projected via paleontological pointers and centered on paleomagnetic statistics (Fairbridge, 1971, 1973). Found formerly was the late Proterozoic period of ice, called the Varangian because of the place of its 1st finding on Varanger Fjord in the district of Finnmark of Norway. The current latitude, almost 100 kilometers north of Arctic Circle, offers no issue in climatology (Frakes, 1979). Ultimately the Eocambrian glaciation was found in West Africa, South America, and South Australia, triggering the idea of Snowball Earth at 1 phase. Plate-tectonic restorations were required, conversely, to smear the actualistic models. Yet former Precambrian tillites have been tracked back in time. The issue of right recognition appears as traced back and the terms diamictite and stony mudstone have been utilized.

2.10. EVENTUALLY: CLIMATIC VARIATIONS

Visher stated: "Geology as the prognostic science might find its origins in the naturalism of the 18th century. The necessity for the naturalistic philosophy was established because of the innovation of the expectedness of the body's motion helpful in determining the array of order all over the entire solar system." The advancement demanded inductive thinking, and, additionally, unflinching theories were required. In the year 1830, Charles Lyell had stated: "Geology is a science which examines the successive variations that have occurred in the inorganic and organic domains of nature; geology probes into the reasons of these variations and the impact which they have applied in adjusting the external structure and surface of the planet" (Figure 2.6).

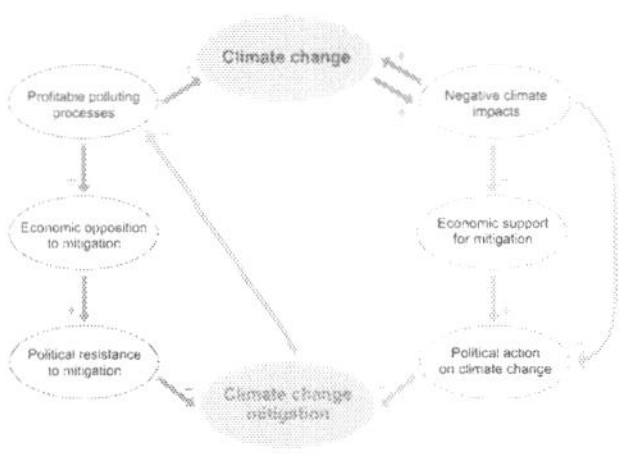

Figure 2.6. Causal loop representation of climate variation.

Source: https://www.researchgate.net/figure/Climate-change-causal-loop-diagram_fig4_286629569.

The universal principles can be applied now to paleoclimatology, given that the interdisciplinary nature of paleoclimatology is comprehended clearly: Meteorology, astronomy, geochemistry, biology, stratigraphy, and sedimentology are all indulged. Generally, the lengthy treatment of paleoclimates was developed by Eugène Dubois, who found the Java Man. Dubois (1893) supposed that the growing Sun was the factor in paleoclimatology, however, the scale of time was a problem. Proof of extraordinary climatic variations that entirely altered central Asia's physical geography and the environment of humans on scales of 100,000 years was discovered in the central Asia (Huntington and Visher, 1922). Initial developments on paleoclimatology comprise Climate Through the Ages (Brooks, 1949). Brooks appropriately arranged the series of possible compelling assistances, with many references. Frakes (1979) and Schwarzbach have significantly extended the record. The real term Palaeoclimatology was used in the text by the paleobotanist, F. Kerner-Marilaun (1930).

2.11. COSMIC VARIABLES

In the 20th century, growing information of outer space provided, for instance, by a Hubble Telescope, generally has increased the ideas of galaxies and mainly of the Milky Way. The main agent of climate modulation that has been recommended is intergalactic dust, along with clouds of gas. The initial recommendation of the latter seems to be in the year 1895 by Calverwell. Cosmic dust in outer space was preferred by Noelke. The 2 astronomers of Cambridge, Lyttleton, and Hoyle, wondering on the prospective for the interstellar matter to adjust weather, have been rigorously condemned for their eccentric ideas (Figure 2.7).

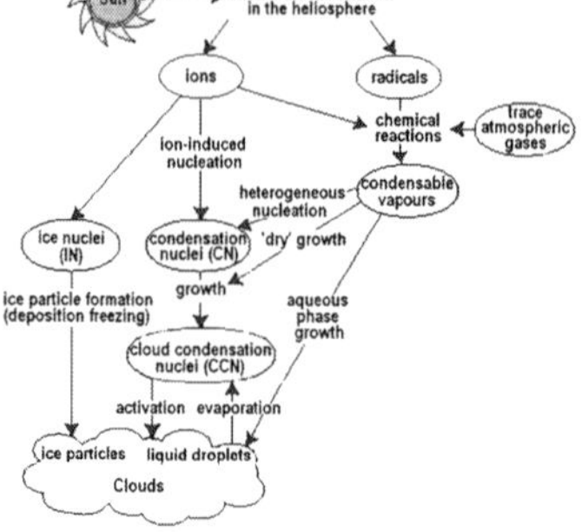

Figure 2.7. Consequences of cosmic rays on climatic changes.

2.12. ELEMENTS IN THE GALACTIC ARMS

The 20th-century investigations of the Milky Way have exhibited that there is a high accumulation of particulate matter implanted in the 2 spiral arms. Moreover, the parts of these arms exhibit a consistent ripple over which the entire Solar system should pass in its rotation around the entire galaxy. The periodicity is nearly 30 Myr, which gives the probable compelling function for evolution spurts (Gould and Eldredge's interrupted developmental version), for which Haggerty and Rampino offered the word Shiva Hypothesis.

2.13. SOLAR RADIATION

Two kinds of energy emissions from the star Sun are particularly appropriate for paleoclimate and climate investigations. Almost more than 99.9% of the energy is in electromagnetic form, spectrum varying from X-rays and gamma through ultraviolet to visible and Infrared to radio waves. The Ultraviolet to near-infrared spectrum is appropriate for the weather (Oort and Peixoto, 1992). It moves in the rectilinear logic with the speed of light (nearly 186,000 miles second), and therefore has its terrestrial influence exploited at the equator of Earth and as far south and north as the 2 tropics, Capricorn, and Cancer, lessening poleward. The flow of heath from within the Earth produced by the nuclear reactions is approximately 3.50 orders of enormousness less as compared to the exogenetic origin (Figure 2.8).

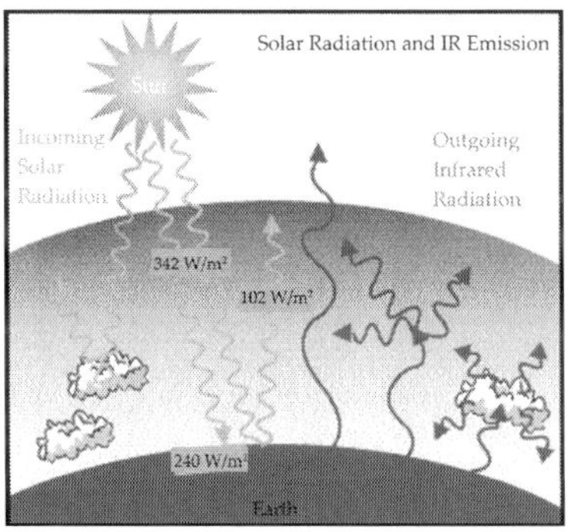

Figure 2.8. Consequences of solar radiation (energy emissions) on climatic changes.

Source: http://www.cotf.edu/ete/modules/climate/GCclimate1.html.

The 2nd solar source is the emission particulate of moving over the solar wind; the flux of plasma, which moves at different speeds but usually requires numerous days in order to reach the Earth. The wind due to the sun transports magnetized particles and as a result, is bounced by the magnetosphere of Earth and directed towards the lower atmosphere. A path all over the atmosphere outcomes in impacts with molecules of O_2, CO_2, and nitrogen, occasioning in the isotopic spalling. The finding of sunspots in the year 1610 by Galileo Galilei opened the path to the idea of the small-scale variables in solar radiation. It occurred that feeble sunspots of the Maunder Minimum throughout the Little Ice Period instantaneously followed the discovery of Galileo, but robust sunspots became noticeable in Korea and China so that some continuousness was sustained.

2.14. ORBITAL DISSIMILARITIES

Even though accurate calculations of orbital features of the motions of Earth, Moon, and Sun have become feasible only in the age of computers, valuable hand computations have generally been accessible for more than 300 years. Association with the terrestrial weather was suggested by Adhémar and established in the year 1891 by James Croll and Robert Ball. An exceptional account of the 19th-century discussions can be accessed in Imbrie and Imbrie (1979) (Figure 2.9).

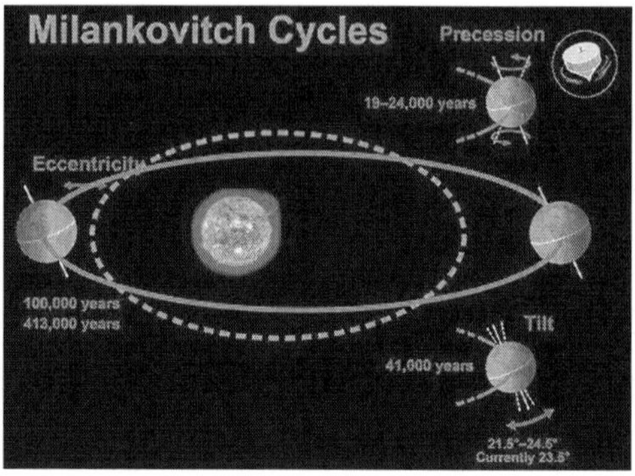

Figure 2.9. Cycles of Milankovitch.

In the 20th century came a mathematician Milutin Milankovitch. Initial debates with geological uses had generally been suggested by Frederick Zeuner and these usually had a universal influence. The reaction of the Establishment was nearly negative. The historians of science will possibly debate over this query for a long time, however, the current writer (R. W. F.), considers that this disagreement was driven by various factors like racial intolerance, professional narrowness, and religious conservatism. The initial history of debate is handled greatly by Imbrie and Imbrie (1979). Consequent developments were observed at various international sessions (Berger et al., 1984, 1992).

The last conference took place in the year 1984 at Lamont Doherty Geological Observatory of the University of Columbia when all of the main experts met under one roof, along with the living folks of Milankovitch family who reached there as guests (Berger et al., 1984). The proof of (a) astronomy was accurate; (b) the timeline of deep sea accumulations with the ^{18}O isotopic paleoclimatic substitutions was irrefutable; (c) the record of deep sea was indistinguishable with the time-series of continents established from glacial-eolian accumulations varying with interglacial deposits of soil with kind sections in China and Moravia (Kukla, 1972); (d) courting the marine coastal accumulations had revealed the eustatic synchrony amongst them and Milankovitch-anticipated cycles of interglacial/glacial as predicted by Maclaren and transformed into the orbital perception by Fairbridge (1961) and Zeuner (1959).

The three main changeable of the Milankovitch standard (efficient terrestrial insolation outcomes from variations in precedence (18.0 to 23.1 kyr), obliquity (41.1 kyr), and the eccentricity (100.2 kyr)), extrapolation for the long term is tough but is aided by the power spectrum examination. Approaches for the later have been refined (Berger, 1995). The values of insolation have been computed for the previous 10 million years (Berger et al., 1992; Berger and Loutre, 1991, 1992).

2.15. EVENTS OF ASTEROID COLLISION

Small asteroids or the large meteor collisions are noticeable from the proof of craters in the semi-dry territory for about 5.0 to 10.5 Myr, however, are hard to track in older sandy valleys and the submarine sites, therefore comprising more than 2/3rd of the surface of Earth (Fairbridge and Shirley, 1997).

The main events of an asteroid, instead, are primarily inferred from secondary proof, like geophysical anomalies, organic extinctions, and debris rings (Rampino, 1998; Rampino and Haggerty, 1996). The instant outcome is the tragic surge of heat, as proved by widespread fragments of charcoal (Anders and Lewis, 1985). This, conversely, is concise but trailed by acid rain and extended universal cooling. The occurrence of 1906 Tunguska was not the main event but the damage of forests was intense. The paleontological record recommends an event of impact is trailed by the climatic worsening of 10^5 to 10^6 years, instigated by the comparatively short-term particle charging of stratosphere, but initiating positive response of particular variety, from the excitation of the plate, tectonic uplift, and the rain shadow dryness to the lowered lines of tree.

The huge destruction at the event of the K/T boundary was reported in the year 1965 by Gramletter, 2 eras before the detections by Alvarez. If an asteroid had fallen in the sea, nearly 1254 billion tons of H_2O would have encumbered the atmosphere and would not just have efficiently extinguished the early fires of forests, nevertheless, its poisonousness should have generally been unfavorable to most kinds of life. The individuals still living were protected by warrens on land, oceanic depths, or subterranean lakes. The Stenothermal kinds of life expired. The Eurythermal kinds had better probabilities of existence. The early tsunami waves touched nearly 1,000 kilometers to Texas from the Chicxulub in Yucatan (Bourgeois et al., 1985), so that the marine littoral biotas would not have enough time to migrate or adapt (Figure 2.10).

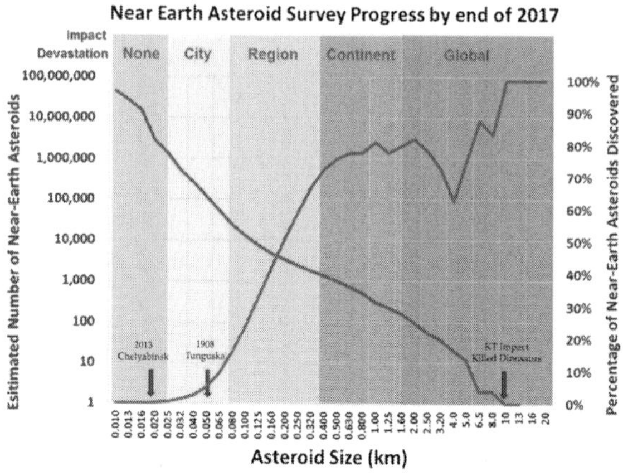

Figure 2.10. Outline of events of asteroid collision.

Source: https://www.extremetech.com/extreme/271953-nasa-asteroid-impact-preparedness-plan.

Following any main event of an impact, there is the general agreement that the nuclear winter would follow. In addition to the effect of climate, the area of impact would be the biological inhospitable surroundings, the biotic vacuum in which novel inhabitants might be very slow to infiltrate. Therefore, the climatic associations linked with the expansion of soil would be considered deferred.

The model of impact for the event of K/T has not been accepted globally and the well-reasoned volcanic substitute (Page and Officer, 1996) has its supporters. On the contrary, the event of Eocene-Oligocene boundary (34.2 Myr) also noticed major destructions and good proof for the asteroid impact (35.0 Myr ago, USA; Poag, 1997), however in this situation the tectonic-eustatic and orogenic model might also provide the reasonable clarification (Mörner, 1984). All of these factors are feasibly interrelated.

2.16. ENDOGENIC GLOBAL HEATING

The flow of heat from the interior of Earth was apparent in the initial times of coal excavating in the 18th century. Validated by subterranean excavation for Au and the other metals in the 19th century, it then became the theme of geophysical curiosity in the 20th century. Edward Bullard invented heat probes for utilization at sea. The basis of inside warming of the Earth could be given by radioactive decay. Throughout early planetesimal in fall, the radioactive elements would produce substantial heat, however, it is just the isotopes that lived long that would produce sources of heat exceeding the planet age. Systematic measurements of the flow of heat around the world reveal energy bases sufficient to activate convective movement and thus drive plate tectonics (Figure 2.11).

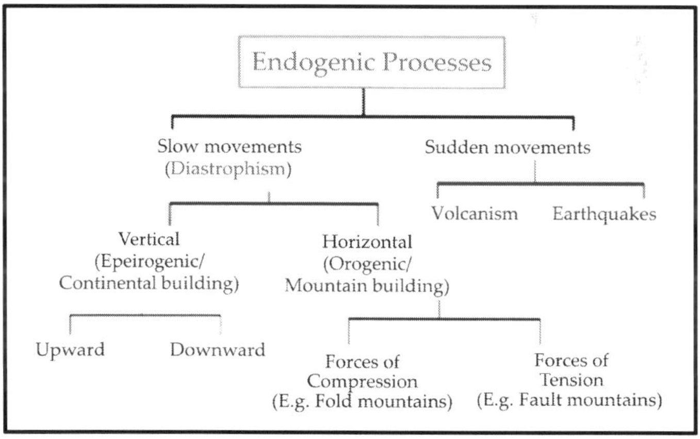

Figure 2.11. Endogenic procedures and their cataloging.

Source: https://www.flexiprep.com/NCERT-Exercise-Solutions/Geography/Class-7/NCERT-Class-7-Geography-Ch-3-Our-Changing-Earth-Part-1.html.

2.17. PLATE TECTONIC AND RELOCATION OF OCEANS AND CONTINENTS

In the 19th century, there prevailed supposition that oceans and continents occupied less or more everlasting positions in the crust of Earth, though marine regressions and transgressions were recognized globally and orogeny was usually an accepted idea. A concise historical analysis is given by Gohau.

In the 20th century came the Alfred Wegener (1880 to 1930) and also the philosophy of continental drift (Wegener, 1966). This theory was acknowledged somewhat positively in South Africa, Australia, India, and South America (Du Toit, 1937), but not much in the Northern Hemisphere (Figure 2.12).

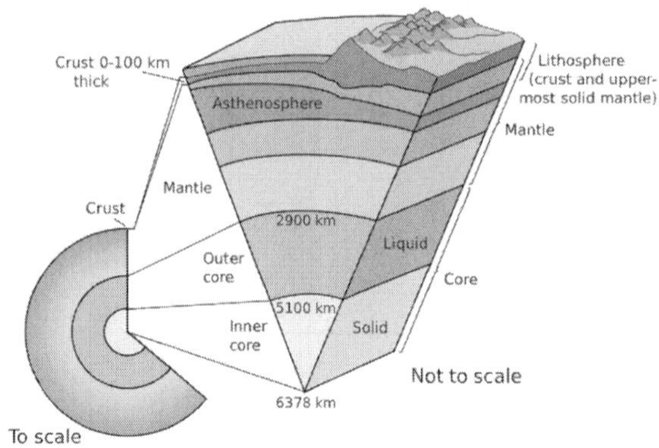

Figure 2.12. Schematic representation of the plate tectonics.

Source: https://www.usgs.gov/media/images/earth-cross-section.

The coastal matching had been observed for South America and Africa by Francis Bacon (1561 to 1626) despite the imperfect plots of the Elizabethan times. Analytic aspects were usually the paleoclimatic signs, found in the 19th century. The seeds having the size of the pea of Permian Glossopteris flora could not be carried all over Gondwanaland by the wind. The Abundantly fossiliferous remains in the seams of coal were interstratified with certainly glacial accumulations on every Gondwana continent. The drift theory of Wegener, critics appealed, though, was (a) incredible in standings of geophysics, however (b) more significantly, it was opposing to religious ideas and old-school viewpoint.

After WW2, novel geophysical instruments altogether with universal marine geological analyses and the Deep-Sea Boring Program triggered overall reconsidering. The ocean paleomagnetic irregularities were the answer to comprehending the vibrant history of oceans. The spreading of Sea-floor unlocked oceanic seaways (Pitman and Heirtzler, 1966; Dietz, 1961, 1977). Mobilism was IN and permanence was OUT. The Philosophy of Plate Tectonic became the geological model of the 20th century. The chronological analysis is given by Marvin (1973).

Whereas plate borders are concluded by seismic belts nowadays and by petrological or structural signs in history, they do not conclude paleogeographic borders and neither do they associated with climatic systems. Precisely it is the previous shoreline that is of specific interest. Nonetheless, these plates are necessary for the reconstruction of continents, and therefore for the restoration of paleoceanograph, but the later needs the eustatic management.

For the periods of Mesozoic-Cenozoic, 2 significant water bodies intensely disturbed and governed worldwide paleoclimatology: (a) the huge Pacific Panthalassa of Wegener, (b) Tethys, which was officially developed by Edvard Suess a long time before the plate tectonic restorations. Over nearly 0.251 billion years, the pacific Panthalassa gradually contracted in size whereas the eastern Pacific was also subducted. Simultaneously, the southwest and west-central Pacific became separated into the island arcs and back-arc basins. The Tethys was associated equatorially with Panthalassa. The association of Tethyan was equatorial, having the oceanic circulations going from east to west. The result of climate was the universal heating system of astonishing continuity and efficiency. Providing that it stayed in place, no global ice period could develop.

As Australia, Africa, and India flowed towards the north, they assisted to develop the incessant Southern Ocean, developing the 2nd worldwide circulation that governs Southern Hemisphere systems of climate nowadays (Fairbridge, 1982). This particular feature has been established in phases over the last 200 million years.

The migrations of a plate in the Cenozoic progressively restricted the previously open Tethys. The current Mediterranean is all of the leftovers of the enormous Tethys (Emiliani, 1992). Ultimately South America flowed on Caribbean and the North American plates. The volcanic belt closed the Panama Isthmus (Kennett, 1982). The mighty Tethys stopped existing as the world-enclosing radiator and the worldwide weather system went into the mode of cooling. The tectonic plate governs for the ice-period foundation were 3 in number: (a) incomplete obstruction of the E-W equatorial currents of the ocean (Fairbridge, 1973); (b) existence of the pole-leaning mass of land (Schwarzbach, 1963); and (c) orographic development (in Europe, North America, and Central Asi; Raymo, 1994; Raymo and Ruddiman, 1992).

Two additional climatic systems narrate to plate-tectonic directs: (a) comparative to continental crustal rifting and the sea-floor scattering and (b) comparative to subduction regions, volcanism, and orogeny.

2.18. OROGRAPHIC REACTION TO EUSTASY AND PLATE TECTONICS

Instances of Quaternary and current-day rifts of continents are observed in East Africa, having the evolution to oceanic settings in the Gulf of Aden and the Red Sea. The Red Sea kind linear beds that were occasionally shut produced evaporating or euxinic furrows, along with locations for the future concentrations of hydrocarbon. In Brazil, the elevated rift shoulders produce the rainforest belt at Santa Catarina.

Continental rifting and the impacts on the vertical state of the crust of Earth can be computed by courting the trace of the earliest shorelines. This presents the intricate issue of eustasy or global shift of the mean sea level, which estimates the geoid (Fairbridge, 1961). This is just an estimation since the real mean sea level can change significantly in time and space. Already stated is the consequence of glaciation on the worldwide level of the sea. Though, the field proof for the previous shoreline, as signified by the earliest beach accumulation, the corrosion bench, surface of the coral ridge, or some other geomorphic characteristics, can imitate the variety of characteristic causative procedures.

Geology professor, Edvard Suess discovered the term eustatic variations of the level of sea (1885 to 1909; 1904 to 1924). Suess observed that these variations were frequent, large-scale variations of global sea level all over geological time. Strangely, though, he did not relate variations with magmatism, glaciation, or tectonism, nevertheless with sedimentation. Therefore, Suess articulated, the slow filling up of the basins of oceans would be accomplished by corrosion and deposit carriage from the elevated masses of land. Ergo, the level of the sea would rise. Even though the term he invented was outstanding, his version was validated to be of least importance due to numerous causes. These generally fall into three categories:

- Tectonoeustasy (functioning by structural distortion of the basins of the ocean, as throughout thumps of the plate tectonic, having an Archimedean effect);
- Glacioeustasy (functioning by variations in the phase of hydrology, water being moved from oceans to the land and then being transformed to glacier ice); and
- Sedimentoeustasy (functioning by deposit-infilling of the valleys of the ocean; the consequence of land corrosion; primarily unidirectional, cycles of sediments interchanging with tectonic lift trailed by cycles of erosion).

2.19. UPLIFT OF MOUNTAIN AND GLOBAL COOLING

Paleogeographic proof exhibits that during the previous 65.1 million years there has been a continuous decline in the mean worldwide surface temperatures. The subduction at dynamic convergent plate limits has caused recurrent surges of orogenic elevation, comprising the formation of a high plateau. High uplifts have caused more snow on mountains, but generally dehydrated circumstances in the rain shadow plateaus. Global and continentality albedo has increased, causing continuous worldwide cooling. Brooks lists many mentions to these procedures and the Tibet lift has produced various publications (Figure 2.13) (Raymo and Ruddiman, 1992).

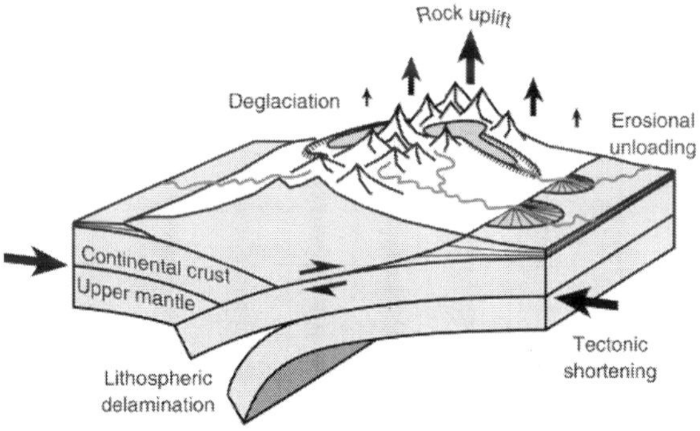

Figure 2.13. Mountain uplift of the European alps.

Source: https://www.nature.com/articles/ncomms13382.

2.20. OCEANIC CIRCULATION

Currently, there exists a general worldwide movement in an ocean, first acknowledged by Stommel, after his research regarding Gulf Stream. It was observed as functioning in the way of the conveyor belts, known as Conveyor (Broecker, 1991). The necessary start identifies the sink of water that becomes the cold, compressed flow of deep-water that heads towards south from Greenland to the Southern Ocean where this flow turns towards east and is then reinforced by cold and high-density water from under the Antarctic Sea ice. The counter hot water flow leads back over the East Indies to the Indian Ocean, nearby Cape of Good Hope, and links with Benguela Current, onward through the equator towards the west in order to link with the Gulf Stream (Figure 2.14).

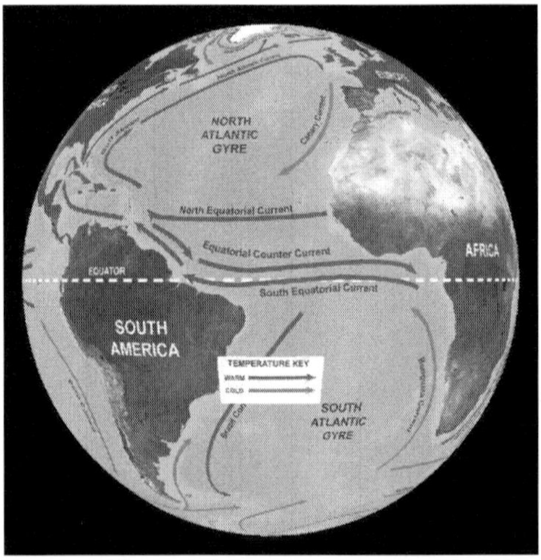

Figure 2.14. Map of the cycles of oceanic circulation.

Source: https://sos.noaa.gov/datasets/ocean-circulation-labeled-currents/.

Geological action along some other submarine edges has possibly played similar roles in the distant past. For instance, Brooks (1949) quotes E. Hull who recommended an Antillean continent in the year 1897 that could have caused the main rebound of the Stream of Gulf. The complete zone has experienced essential reorganizations as the Cretaceous, ultimately causing the hindering of Tethys and recurrent closing and opening of Panama Strait. The last shutting in Pliocene allowed 2-way movements by the animals of land and obstructed the marine networks (Stewart, 1975).

According to worldwide heating of the Cretaceous times, T. C. Chamberlin recommended that at the crucial point, augmented evaporation in lower latitudes might revert the current movement. The worldwide salinity sink could be equatorial. The upsurge in polar zones would normally keep them everlastingly hot.

2.21. SEA ICE AND THE GLACIER VARIATION

The existence of the sea ice in the winter season is considered to be a basis of the designation of Iceland later on developed by Scandinavian refugees after the 8th century, around the Charlemagne time and the Roman Empire. It was an era of extraordinary warmness climatically: the Little Climatic Optimal or the Medieval Warm Era (Lamb, 1977). The rich plant life thrived at the time in Spitsbergen. It persisted until around AD. 1301 which labeled the commencement of the changing sequence of cold interims, the Little Ice Period, which continued till around AD. 1726. The Vikings could not have been more privileged in their settlement journeys, since there is proof that the westerly hurricane paths in summer seasons inclined to head over Spitzbergen and Greenland. Therefore, there were often easterlies in order to fill up the sails. For the short time, AD. 1001, they had the base at Anse-les-Meadows, the 1st European colonization. The Little Ice Period turned the circumstances against Vikings. In spite of the hot cycles, the overall inclination was shifted from bad to even poorer. The colony of South Greenland progressively died and the final Viking deceased nearly AD. 1400 (Figure 2.15) (McGovern, 2000).

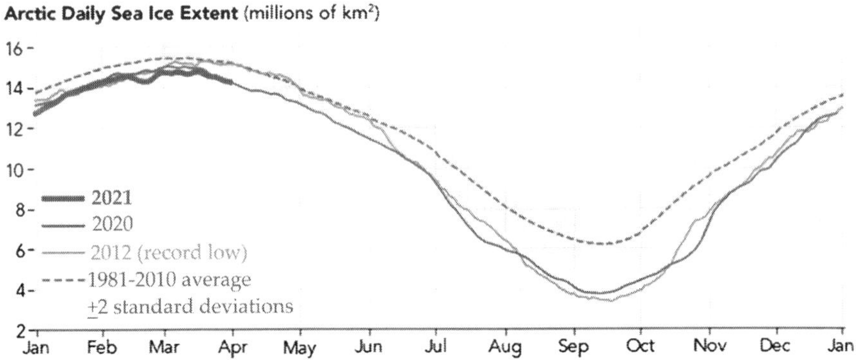

Figure 2.15. A yearly discrepancy of the arctic sea-ice amount over the years.

Source: https://earthobservatory.nasa.gov/world-of-change/sea-ice-arctic.

Sea-ice develops sometimes today in the winter season on the coast of Iceland when the language of the Gulf Stream weakens. Main and extended intrusions were documented throughout the Little Ice Period (Grove, 1988). The greatest rigorousness, conversely, was observed throughout the 17th, 18th, and 19th centuries. Lamb considered sea ice, over the albedo, as the key aspect in high latitude cooling.

Currently, the largest glacier of Iceland, Vatnajökull, covers nearly 8,301 square kilometers, and there exist 6 smaller ones. The initial research of mass balance on the Vatnajökull was carried out on the mutual Swedish-Icelandic voyage, carried out by Jon Eythorsson and Hans Ahlmann. Some confusion existed earlier regarding the glacier's response to climatic compelling, but Ahlmann (1938) showed that in this circumstance cooling of climate was the leading management. The philosophy of ice eras by Ewing-Donn (Ewing and Donn, 1956), conversely, insisted that rain was the main key. They offered rather that throughout interglacial warming of the oceans of the globe, the Arctic Ocean was essentially ice-free; therefore, all of the glaciers might enlarge and spread. Though, the Arctic Ocean does not seem to generally have been completely free of ice for many 1,000 years. In the 2nd half of the 20th century, there generally has been the drawback of mountain glaciers and sea ice (Kerr, 2006).

2.22. VARIATIONS IN ATMOSPHERIC COMPOSITION

In initial Precambrian times, the geological proof discloses secular and episodic variations in the atmosphere of Earth. These variations have been enforced by several organizations, that have been subjected to advanced alteration. Not all the aspects are understood, however, one undeniable point arises from the geologic document: the incessant presence of water in the fluid state. It trails that the Earth, Moon, and Sun connections and the planetary orbits should have been preserved in the active balance on several million years scales of time (Shirley and Fairbridge, 1997); moreover, it trails that cyclical or secular changes in the solar radiation should be well-adjusted by dominant mechanisms of response that have preserved the quasi-stable weather and averted the mean temperature of Earth from ever experiencing the runaway heating or the runaway icehouse condition, even though debates for the state of snowball have been given.

Secular variations of the geological time are revealed by sedimentological-paleontologic-stratigraphic records and its consecutive biogeochemical insurgencies. The variations are briefed as (Oliver and Fairbridge, 1987):

- 1st life, in CO_2 or CH_4 environment (4.0 billion years);
- Mid-Precambrian verge of oxygen (2.4 billion years);
- Cambrian/Ecocambrain $CaCO_3$-exoskeletal life (0.541 billion years);
- Late Paleozoic Great Coal Period storage of carbon (0.351 to 0.291 billion years);
- Cretaceous-Paleocene storage of carbonate (0.151 to 0.501 billion years).

Inferior, cyclical variations have increased or slowed the secular inclinations. The periodicities of cyclical variations vary from variations in galactic arms at around 0.3 billion years (Haggerty and Rampino, 1996) to Milankovitch parameters of orbit, as observed earlier (Figure 2.16).

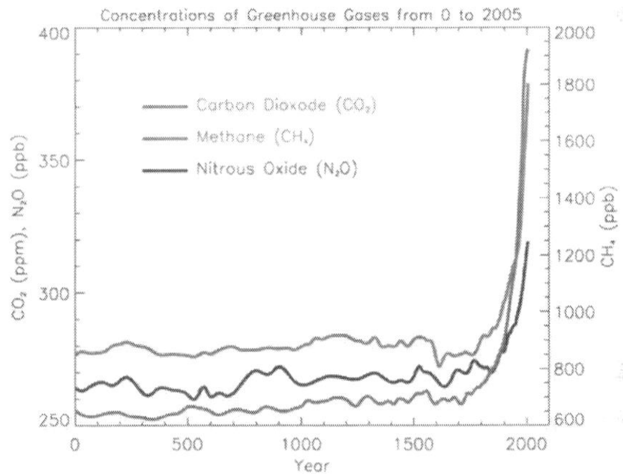

Figure 2.16. Fluctuations in the concentration of the greenhouse gases from zero to 2005 A.D.

Source: https://www.e-education.psu.edu/meteo300/node/606.

In addition to biogeochemical procedures that primarily rely on climatically governed surroundings, there exist asteroid collisions, and tectonic and volcanological procedures that are non-cyclical largely, or just slightly. Volcanoes, reliant on the magmatic features, emit water vapor, SO_2, CH_4, and CO_2. Climatically significant is the point that water, CO_2, and methane are conservatory gases and they incline to be ecologically heating, contrary to the volcanic aerosol silicon dioxide, which is the cooling agent.

CH_4 is an efficient greenhouse gas (GHG). Methane is created by a broad variety of organisms, from ungulate cattle and termites to wetland bacteria.

Generally, these sources incline towards the tropical climatic partiality, but the tundra wetlands are too significant.

The conservatory gases not just disturb weather but also affect worldwide hydrology (Herschy and Fairbridge, 1998) and the eustatic variations of the level of sea (Fairbridge, 1961). The 2 are related, for instance, throughout the main cycle of cold when the level of the sea is low and the biotic activity is decreased worldwide. Throughout glaciations, the low temperatures usually decrease metabolic rates, H_2O is limited and the deserts are common; atmospheric water, carbon dioxide, and methane levels are all decreased. Also associated with the hydrology-eustasy connection are the biotas of soil. The soils of the world are usually ignored by climatologists and geologists but the atmospheric part is of main significance. As stated, Erhart's biostatic circumstance is the long-term, stable, hot, high eustatic state, having steady carbon dioxide (Erhart, 1956; Fairbridge and Finkl, 1980). This is distinguished by Erhart's rhexistatic circumstances that are striking by sharp occasions of cooling, having eustatic fall causing widespread erosion, consequential in transmissions of the iron oxides to deposited storing. Natural deforestation because of desertification and glacial advance shifts carbon dioxide to the atmosphere so that the pendulum swings gruffly into the warming mode again.

In the 21st century, a substantial focus is being laid on the problem of greenhouse heating. This circumstance has generally been brought by anthropogenic movement, mainly expressed by growing levels of carbon dioxide, silicon dioxide, and methane, primarily subsequent upon deforestation and also the burning of hydrocarbons. This activity started with the evolution of Homo sapiens, beginning around 4001 BC. with charcoal-centered founding (Ruddiman, 2005). Deforestation extended with the requirement for augmented pasturage and the planting of muesli crops (Lamb, 1977).

2.23. THE MOON, LUNAR-SOLAR TIDES, AND ASSOCIATED SOLAR, TERRESTRIAL, AND PLANETARY CYCLES

The only satellite of Earth, the Moon possesses an exceptionally large ratio of mass (1:80), and the 2 act nearly as identical planets, rotating around the usual center of mass situated nearly 4,801 kilometers from the geo center. The orbital plane of the Moon is inclined at the angle of 510° to the Earth's orbital plane, and the exact spot where they cross each other is known as

Node. The Moon's elliptical orbit at the perigee radius is nearly 365,000 kilometers and at apogee around 410,000 kilometers (Shirley and Fairbridge, 1997). This causes cyclical variations in climate-associated periodicities and tides (Currie, 1995). The main weather impact in the 1860-year nodal cycle, perceived in temperature and precipitation time series.

The tides' rising force of the Moon is 2.18 times as compared to that of the Sun and the subsequent is the lunar-solar tides, which produces the oceanic tidal waves that circle the world each day, although altered by the land obstructions. Over the time of almost a fortnight, the range changes from spring to the neap tide, dependent on the comparative alignment of Moon and Sun. Augmentation of the range happens at perihelion when the Earth is nearest to the star Sun in the northern hemisphere. Whereas it occurs in July in the southern hemisphere. The tide is quite high at perigee and higher at the syzygy. The close line ups take place at intervals of 558.5 years and 93.07 years. The multiples of 1860 year nodal cycles are observed in the long climatic substitutions (Currie, 1995) along with the short-term harmonics, which associate with El Ninos (Fairbridge, 1990).

Otto Pettersson (1912, 1930) discovered that the 1860-year nodal cycle produced the internal wave in the North Sea at the border of intensifying offshore salty waters and to somewhat freshwater outflow from the Baltic Sea. As the wave advanced the coast near the gradually flimsy North Sea, this internal wave becomes steeper and larger. At the tidal station on Gulmar Fjord usually, it can reach 24.5 meters and trigger precarious currents. The limits are approaches at 93.07 years, the 5 cycles reappear at the same season, and similarly 15,317.65 hours. Pettersson found that intensifying super tides brought the young herrings flood that thrived and brought substantial revenues to the fishermen of the Hanseatic League (Figure 2.17).

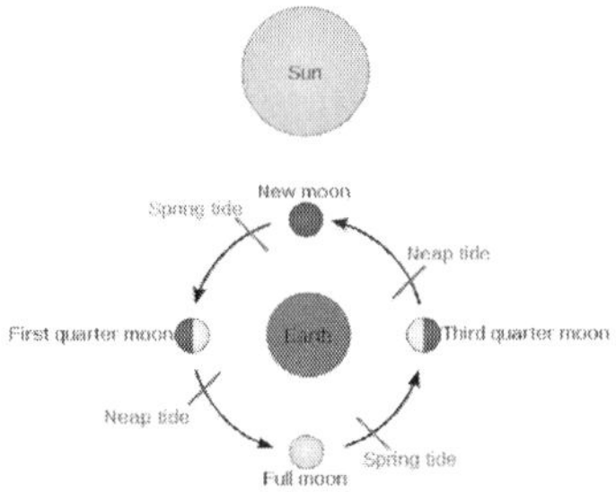

Figure 2.17. Processes of the lunar-solar tides.

Source: https://tambonthongchai.com/2018/08/05/tidalcyclesbiblio/.

The year 1912 brought with it the highest anticipated astronomical tides of the 20th century. The super tides lead to moving glacier snout calving in Greenland and in the year 1912 a remarkable stream of floating ices came in the path of the Titanic (Wood, 2001). The warming waters and buoyancy effect initiates the main launching of the iceberg and not the cycles of cooling as supposed sometimes. The lunar cycles might also be indulged in Heinrich events that spot the last 80,000 years in North Atlantic. The staircase of 186 isostatically appeared beach ridges on the Bay of Hudson at about 44.5-year periodicity recommend storm cycles of the synergetic lunar-solar forcing (Fairbridge and Hillaire-Marcel, 1977), with the peaks at 316 years. Analogous but less-comprehensive sequences take place in another place in northern Norway and the Arctic (Fletcher et al., 1993).

The nodical cycle of 18.60 years marks changes in the declination of the Moon (Wood, 2001). This outcome in the zenith position of the Moon shifting over 2,846 or 15,689-kilometers S or N in 9.3 years, recurring at about a similar season in the 317-year lunar-solar period. When the zenith position moves south or north it has the impact of decreasing or increasing current velocities of meridional tides and also the heat transport of these tides. This effect is obvious in the meridional currents such as Gulf Stream, which reveal a > 1°C fall or rise in associated surface temperatures. Maksimov (Lamb, 1977), a Russian oceanographer, discovered that it also disturbed the length of sea-ice in the Russian Arctic and the Barents Sea. It might thus disturb global albedo.

It appears apparent that the Moon might also play the climactic part in the equatorial belt. Mainly important is the blockage made by East Indies due to their shallow seas and narrow straits, to some extent hindering the flow of returns of the Conveyor system along with east to the west tidal wave as it flows from the Pacific Ocean to the Indian Ocean. The archipelago takes an exclusive place on this planet; numerous volcanic cones focus on the intensifying warm and humid air, serving it in the diurnal pulses to the stratosphere. The latter is open to modulation by the arriving solar radiations, the pulse being the quasi-biennial oscillation (QBO) of solar radiations at 2.162 years, which is the characteristic terrestrial climate cycle (Schove, 1983). Conferring to Lamb (1977), QBO (2.162 years) is the global climate cycle over the annual one. QBO is the most omnipresent solar radiation period.

2.24. VOLCANIC ERUPTIONS AND THE DESERT DUST

Main volcanic eruptions of the historical period, like Krakatoa (1883); and Tambora (1815), have been linked with various years of worldwide cooling (Lamb, 1977). The atmospheric consequences of the Krakatoa eruption and the clouds of dust were discovered in the year 1901 by the Sarasin brothers, and the general theory was given in the year 1913 by Humphreys. The databank was strengthened greatly by the global collection of more than 1,000 years of historical and geological records (Sigurdsson et al., 2000). Bryson (1997) recommended that precipitation variations and large temperature, particularly in the mid-Holocene, imitated planetary-scale triggering of volcanic dust. From the Indian environmental signs, he determined that the trends of cooling were noticed by the firming of a circum-polar vortex and the shift of jet stream towards the south. The shift is mainly clear in northern India and the Mediterranean, which are visited by abundant winter rains. In the belt of monsoon, though, the rains of summer are strengthened. A conference sponsored by NATO in Turkey (Dalfes et al., 1997) well-thought-out about these questions, particularly in association with the conventional failure of progressive civilizations there in the 3rd millennium BC. The dust of Lee-desert was reported first by Charles Darwin in the year 1832 on the journey of H.M.S. Beagle off the shore of West Africa. Distinguishing hints of microfossils from Lake Chad have been tracked towards the west till the Florida, Caribbean, and New York City. The cores of the Caribbean deep-sea (Emiliani, 1966) and the ice cores of the Antarctic (Petit et al., 1999; EPICA, 2004) exhibited that, throughout the last 7 glacial cycles, a load of a rose as the temperatures of water fell (Figure 2.18).

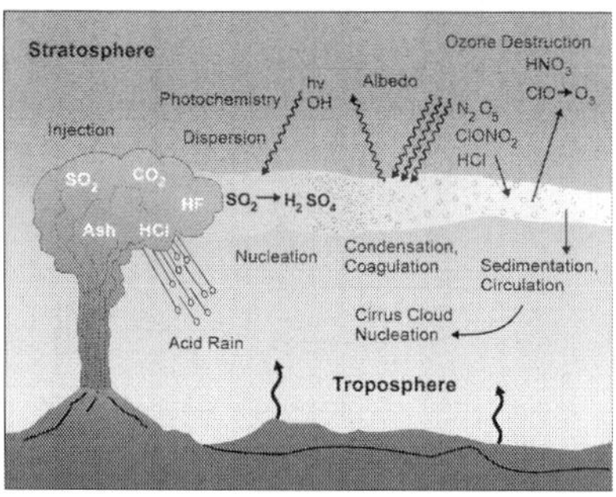

Figure 2.18. Effects of volcanic eruptions on climate fluctuation.

Source: https://www.usgs.gov/natural-hazards/volcano-hazards/volcanoes-can-affect-climate.

Throughout the last glacial phase of the Quaternary ice period, which attained its maximum nearly 17,000 to 26,000 years ago, lower levels of sea triggered upsurged continentality and the colder temperatures of the ocean to reduced evaporation. There was extensive dryness that prolonged from current Mediterranean belts and reached the equator (Fairbridge, 1964).

2.25. PALEOSOLS AND LOESS

Loess is a German term, with a similar origin as the word lose, for the friable soil. It is a climatically particular windblown silt, observed nowadays in the lower area of the Yukon delta, where the strong winds disturb the area with no trees of plaited stream channels and arid alluvial tracts. Though enormously limited in extent nowadays, throughout the waning and maximum phases of the glacial phases, distribution of loess was almost all around the world. Traditional loess is marginally calcareous silt that spreads across eastern and Western Europe into Mongolia, southern Russia, and northern China. It was initially positively accredited to the eolian movement by Baron von Richthofen (1877 to 1885), but this term has long remained debatable. Throughout the warm cycles, surviving loess produces the distinguishing red paleosol (Figure 2.19).

Figure 2.19. Picture presentation loess and paleosols.

Source: https://www.researchgate.net/figure/Photograph-of-loess-and-paleosols-representing-the-last-interglacial-glacial-cycle_fig4_290301792.

The paleosol/loess cycles in Moravia are disclosed in a remarkable series 10s of m (meter) thick in the brick clay pit at Cerveny Vlchy. George Kukla gathered sensibly slanting specimens for the paleomagnetic measurements, which gave the chronologic key. The uppermost region of the series was regular, concerned with the present North magnetic pole, and the lower most region of the series was upturned, therefore leaning to the south. Kukla matched his terrestrial climatic series with $d^{18}O$ paleotemperature fluctuations in the sediment cores of deep-sea at the University of Columbia. The equal number of the climatic cycles down to the boundary of Bruhnes-Matuyama was discovered in the sequences (Kukla, 1972). It revealed over whelming that weather fluctuations in mid-oceans are coincident with the ones in mid continents; moreover, in the later, the temperature fluctuations can be measured with substitute data. Utilizing the Kukla-invented example, the huge loess accumulations of northern China have generally been methodically measured, revealing not just the ideal association with the central European series and deep-sea, but yet initial phases, comprising the Pleistocene and Pliocene.

2.26. EOLIANITE AND EOLIZATION

The termeolization was discovered by André Cailleux, a French sedimentologist. Numerous sandstones and sands of the Paris valley were eolian or mainly oceanic, but throughout negative eustatic fluctuations and open to the action of wind throughout long eras of dryness, they became rearranged, therefore, eolization. In the hot shoreline belts where fragmented corals, molluscan husks, and carbonate foraminifera usually establish the major sands of the beach, short stages of the reduced level of sea support the accumulation of carbonate shore banks. The fluctuations of weather throughout the Holocene recurrently caused eustatic extraction by up to one meter, which caused enormous shore-parallel banks, in places surpassing 20 meters in height. The winds with continued high velocities might develop sporadic blowouts at the gaps in shore-parallel bank belt, therefore serving linear heaps that spread inland, one kilometer. Such kind of high velocities was preferred in specific areas where the usual westerly flow of air might be augmented by the sea-breeze influence distinctive of the limits of subtropical anticyclones. The precipitation throughout the cool intervals triggered an incomplete way out of carbonate dunes sand to combine it into the eolianite or bank rock. Ice-period winds were certainly stronger as compared to the fundamental winds in similar localities nowadays (Jelbring, 1998).

The necessary synergy amongst supply of carbonate of sand continued westerlies and the subtropical breeze of the sea was given during the Quaternary era sometimes conforming to the recurrent negative level of sea fluctuations with augmented wind speeds but also where plentiful carbonate supplies of sand were accessible, suggesting comparatively hot ocean waters. Perfect sites for the highly particular paleoclimatic signs were continental shores and oceanic islands between 35° and 25° S and N, for instance in the Bermuda, and Bahamas to Palestine, Egypt, Lebanon, and Israel, South Africa, India, Victoria, South Australia, and Hawaii.

During the Napoleonic battles, Lt. Nelson, a British naval officer, observed the inquisitive bank features transformed to the firm rock in Bermuda, which gives an outstanding, simply sawn construction stones (Sayles, 1931). In the year 1791, in Australia, Capt. George Vancouver, again a British naval officer, had also remarked on them (Fairbridge and Teichert, 1948). Charles Darwin also observed eolianite on St. Helena and Ascension islands. He stated that since there were fundamentally no beaches on the islands nowadays, the dunes of sand should have generally been ballooned to the higher slopes throughout times when the level of the sea was lesser and the winds were quite stronger. Of course, that analysis was before the glacial philosophy and the process of glacioeustasy (Fairbridge, 1961).

Eolianites and eolization are therefore 2 of the irrefutable paleoclimate signs. Eolization demands severe dryness and continued winds with high velocities where there exists an abundant supply of pebbles, sand, or fragments of rock. Eolianites are certain, being limited to the Mediterranean latitudes 25° to 35° North and South, to be exact, to the littoral belts where the waters of warm offshore favor an abundant carbonate-shelly fauna.

2.27. LEVELS OF THE LAKE AND FLUVIAL SIGNS

The lakes and the levels of lakes establish a convincing found atrion for previous weather assessment. They might behave as the quantifiable rain-gauge however this needs to be understood with care due to the balance of evaporation and rainfall and some other aspects. The efficient management of lake level timeline has been given by Grove and Street (1979), with Australia and Africa as the comprehensible grouping (Rognon and Williams, 1977). The North American pluvial circumstances were fundamentally independent (Figure 2.20) (Nicholson and Flohn, 1980; Bradley, 1985).

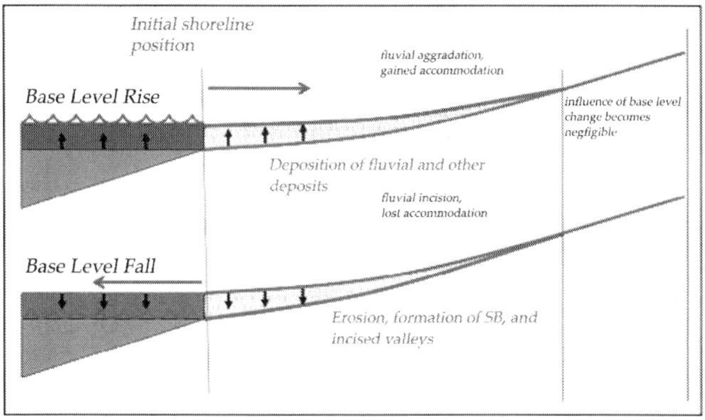

Figure 2.20. Profile of longitudinal deposition displaying a progressive downstream reduction in the slope gradient. Loss and gain of lodging space for gathering of fluvial accumulations during an increase in the base level and corrosion and creation of notched valleys during the fall in base levels.

Source: https://www.sciencedirect.com/topics/earth-and-planetary-sciences/fluvial-deposit.

Factually, the flood of Nile is documented in a perfect manner, beginning with Herodotus around 451 BC., who as a librarian in the Alexandria was certainly accustomed to the pyramids and the delta. The upper sequence was known badly but Eratosthenes had made a sensible plot of the Nile, although the course of White Nile was yet unclear. The main yearly flood was acknowledged as arriving from the Blue Nile, whereas the frail flood of winter was arriving from White Nile. The previous is served by the rainfalls of monsoon in Ethiopia, which decline when the pressure is high in Cairo, with respect to the computations by Henry Lyons in the year 1905 (Brooks, 1926/1949). Brooks perceived that a similar association occurs all over the belt of the Asian monsoon. The floods of the White Nile, conversely, seem to reveal the overall equatorial movement. The hereditary history and the prehistory of the Nile are exclusive among the rivers in that the Nile has offered investigators with an instead safe method of dates centered on archaeology (Wendorf and Schild, 1976) and the radiocarbon-old geomorphology and sedimentology (Williams and Faure, 1980).

Accurate computations of the floods of the Nile were made ever since the time of Ptolemy pharaohs, for a better reason than the wealth of Egypt was dependent on the agriculture retained by the flood of the Nile. The complete records are available now. Around 641–1480 AD but they were just irregular from then till 1830. Brooks cites a complete article by Omar Tousoun on this subject, which defines pre-Islamic floods of Nile simply as excellent, strong, or week for numerous eras since the 5th century BC. Current power spectrum studies have revealed the 11 to 22-year cycle of sunspot and also the 1860-year lunar cycle (El Fandy et al., 1994) but the variations of long term are obvious. The identical periodicities were attained from data of floods from China (Currie and Fairbridge, 1985).

None of the long-term accurate documentation of other main rivers is accessible, but numerous written records prevail for the European rivers (Danube, Rhine). During the previous century, hydrometric of all main rivers have generally been computed but the interval has overlapped with the global eruption of dam construction for control of flood, hydroelectric, urban utilization, and irrigation; therefore, the records have been affected by anthropogenic influences. For an overall management, The Encyclopedia of Hydrology and Water Resources (Fairbridge and Herschy, 1998) might be useful.

REFERENCES

1. Ahlmann, H. W., (1938). *Land of Ice and Fire* (Vol. 1, p. 271). London: Kegan Paul.
2. Albritton, C. C. Jr., (1975). *Philosophy of Geohistory* (Vol. 12, p. 386). Stroudsburg: Dowden, Hutchinson & Ross.
3. Alvarez, W., (1986). *Toward a Theory of Impact Crises* (Vol. 67, pp. 649, 658). EOS.
4. Berger, A., & Loutre, M. F., (1991). Insolation values for the climate of the last 10 million years. *Quaternary Sci. Rev., 10*, 297–317.
5. Berger, A., & Loutre, M. F., (1992). Astronomical solutions for paleoclimate studies over the last 3 million years. *Earth Planet. Sci. Lett., 111*, 369–382.
6. Berger, A., (1995). Modelling the astronomical theory of paleoclimates. In: Finkl, C. W., (ed.), *Cycles, J. Coast. Res.* (Vol. 17, pp. 355–362) (special issue).
7. Berger, A., Imbrie, J., Hays, J., Kukla, G., & Saltzman, B., (1984). *Milankovitch and Climate: Understanding the Response to Astronomical Forcing* (Vol. 1, p. 895). Dordrecht: Reidel.
8. Berger, A., Loutre, M. F., & Laskar, J., (1992). Stability of the astronomical frequencies over the Earth's history for paleoclimatic studies. *Science, 255*, 560–566.
9. Bond, G., Kromer, B., Beer, J., Muscheler, R., Evans, M. N., Showers, W., Hoffmann, S., Lotti-Bond, R., Hadjas, I., & Bonani, G., (2001). Persistent solar influence on North Atlantic climate during the Holocene. *Science, 294*, 2130–2136.
10. Bourgeois, J., et al., (1985). A tsunami deposit at the Cretaceous-Tertiary boundary in Texas. *Science, 241*, 567–570.
11. Bradley, R. S., (1985). *Quaternary Paleoclimatology: Methods of Paleoclimatic Reconstruction* (Vol. 1, pp. 1–24). London: Allen and Unwin.
12. Bradley, R. S., (1999). Climatic variability in sixteenth-century Europe and its social dimension: Preface. *Clim. Change*, *43*(1), 1–2.
13. Bralower, T. J., (2002). Evidence of surface water oligotrophy during the Paleocene-Eocene thermal maximum: Nannofossil assemblage data from ocean drilling program site 690, Maud Rise, Weddell Sea. *Paleoceanography*, *17*(2), 1023, doi: 10.1029/(2001)PA000662.

14. Broecker, W. S., (1991). The great ocean conveyor. *Oceanography, 4*, 79–89.
15. Brooks, E. P., (1926/1949). *Climate Through the Ages London* (Vol. 1, p. 395). Ernest Bonn.
16. Bryson, R. R., (1997). Proxy indications of Holocene winter rains in southwest Asia compared with simulated rainfall. In Dalfes, H. N., Kukla, G., & Weiss, H., (eds.), *Third Millennium BC Climate Change and Old-World Collapse* (Vol. 145, p. 465). NATO ASI Series, Berlin, and Heidelberg: Springer.
17. Caillon, N., et al., (2003). Timing of atmospheric CO_2 and Antarctic temperature changes across *Termination III. Science, 299*, 1728–1731.
18. Cerling, T. E., (1991). Carbon dioxide in the atmosphere: Evidence from Cenozoic and Mesozoic paleosols. *Am. J. Sci., 291*, 377–400.
19. Chamberlin, T. C., (1897). A group of hypotheses bearing on climate change. *J. Geol., 5*, 653–683.
20. Chandler, M. A., Rind, D., & Thompson, R. S., (1994). Joint investigations of the middle Pliocene climate II: GISS GCM Northern Hemisphere results. *Global Planet. Change, 9*, 197–219.
21. Chappellaz, J., et al., (1997). Changes in the atmospheric CH_4 gradient between Greenland and Antarctica during the Holocene. *J. Geophys. Res., 102*(D13), 15987–15997.
22. Croll, J., (1875/1885). *Climate and Time, in Their Geological Relations: A Theory of Secular Changes of the Earth's Climate* (Vol. 2, p. 577). London and New York: Appleton.
23. Crowell, J. C., (1999). Re-Mesozoic ice ages. *Geol. Soc. Am. Mem., 192*, 106.
24. Currie, R. G., & Fairbridge, R. W., (1985). Periodic 18.6-year and cyclic 11-year induced drought and flood in Northeastern China and some global implications. *Quaternary Sci. Rev., 4*, 109–134.
25. Currie, R. G., (1995). Variance contribution of M_n and S_c signals to Nile River data over a 30–8-year bandwidth. *J. Coast. Res., 17*, 29–38.
26. Dalfes, H. N., Kukla, G., & Weiss, H., (1997). *Third Millennium BC Climate Change and Old World Collapse* (Vol. 149, pp. 1–32). NATO ASI Series, Berlin and Heidelberg: Springer.
27. Dietz, R. S., (1961). Continent and ocean basin evolution by spreading of the sea-floor. *Nature, 190,* 854–857.

28. Dietz, R. S., (1977). Plate tectonics: A revolution in geology and geophysics. *Tectonophysics, 38*, 1–6.
29. Dlugokencky, E. J., Steele, L. P., Lang, P. M., & Masarie, K. A., (1994). The growth rate and distribution of atmospheric methane. *J. Geophys. Res.*, *99*, 17021–17043.
30. Dowsett, H. J., & Cronin, T. M., (1990). High eustatic sea level during the middle Pliocene: Evidence from southeastern U. S. Atlantic coastal plain. *Geology*, *18*, 435–438.
31. Dowsett, H. J., Chandler, M. A., Cronin, T. M., & Dwyer, G. S., (2005). Middle Pliocene Sea surface temperature variability. *Paleoceanography*, *20*(2), doi: 10.1029/(2005)PA001133.
32. Dowsett, H., Barron, J., & Poore, R., (1996). Middle Pliocene Sea surface temperatures: A global reconstruction. *Mar. Micropaleontol.*, *27*(1–4), 13–25.
33. Du Toit, A. L., (1937). *Our Wandering Continents* (Vol. 1, p. 366). Edinburgh: Oliver and Boyd.
34. Dubois, M. E. F. T., (1893). *The Climates of the Geological Past and Their Relationship to the Evolutionary History of the Sun* (Vol. 1, pp. 1–23). Leipzig.
35. Ekholm, N. G., & Arrhenius, S., (1898). About the influence of the moon on the northern lights and thunderstorms. *Kungl. Boktryckeriet PA Norstedt & Söner, 31*(2), 77–156.
36. El Fandy, M. G., et al., (1994). Time series models adaptable for forecasting Nile floods and Ethiopian rainfalls. *Bull. Am. Met. Soc.*, *75*(1), 83–94.
37. Eldredge, N., & Gould, S. J., (1972). Punctuated equilibria: An alternative to phyletic gradualism. In: Schopf, T. J. M., (eds.), *Models of Paleogeology* (Vol. 1, pp. 82–115). San Francisco: Freeman Cooper.
38. Emiliani, C., (1966). Isotope paleotemperatures. *Science, 154*, 851–857.
39. Emiliani, C., (1992). *Planet Earth* (Vol. 1, p. 718). Cambridge, UK: University Press.
40. Enting, I. G., (1987). On the use of smoothing splines to filter CO_2 data. *J. Geophys. Res.*, *92*, 10977–10984.
41. EPICA Community Members, (2004). Eight glacial cycles from an Antarctic ice core. *Nature*, *429*(6992), 623–628.

42. EPICA, (2004). Eight glacial cycles from an Antarctic ice core (benchmark papers in geology: Vol. 13). *Nature, 429*, 623–628.
43. Erhart, H., (1956). The Genesis of Soils-Outline of a Geological Theory: Biostasis and Rhexistasis (Vol. 1, p. 90) Paris: Masson.
44. Etheridge, D. M., et al., (1996). Natural and anthropogenic changes in atmospheric CO_2 over the last 1000 years from air in Antarctic ice and firn. *J. Geophys. Res.*, *101*(D2), 4115–4128.
45. Ewing, M., & Donn, W. L., (1956). A theory of ice ages. *Science, 123*, 1061–1066.
46. Ewing, M., & Donn, W. L., (1958). A theory of ice ages, II. The theory that certain local conditions caused Pleistocene glaciations discussed further. *Science, 127*, 1159–1162.
47. Fairbanks, R. G., (1989). A 17,000-year glacio-eustatic sea level record: Influence of glacial melting rates on the Younger Dryas event and deep-ocean circulation. *Paleoceanography, 342*, 637–642.
48. Fairbridge, R. W., & Finkl, C. W. Jr., (1980). Cratonic erosional unconformities and peneplains. *J. Geol., 88*, 69–86.
49. Fairbridge, R. W., & Hillaire-Marcel, C., (1977). An 8,000-year paleoclimatic record of the 45-yr "double-hale" solar cycle. *Nature, 268*, 413–416.
50. Fairbridge, R. W., & Sanders, J. E., (1987). The Sun's orbit, A.D. 750–2050: Basis for new perspectives on planetary dynamics and EarthMoon linkage. In: Rampino, M. R., Sanders, J. E., Newman, W. S., & Königsson, L. K., (eds.), *Climate: History, Periodicity, and Predictability* (Vol. 1, pp. 446–471). New York: Van Nostrand Reinhold.
51. Fairbridge, R. W., & Teichert, C., (1948). The Low Isles of the great barrier reef: A new analysis. *Geogr. J., 111*, 67–88.
52. Fairbridge, R. W., (1961). Eustatic changes in sea level. In: Ahrens, L. H., et al., (eds.), *Physics and Chemistry of the Earth* (Vol. 4, pp. 99–185). London: Pergamon Press.
53. Fairbridge, R. W., (1962). New radiocarbon dates of Nile sediments. *Nature, 196*, 108–110.
54. Fairbridge, R. W., (1964). The importance of limestone and its Ca/Mg content to paleoclimatology. In: Nairn, A. E. M., (ed.), *Problems in Paleoclimatology* (Vol. 1, pp. 431–530). New York: Wiley.

55. Fairbridge, R. W., (1966). *The Encyclopedia of Oceanography* (Vol. 1, p. 1021). New York: Reinhold.

56. Fairbridge, R. W., (1967). Carbonate rocks and Paleoclimatology in the Biogeo-chemical history of the planet. In: *Developments in Sedimentology* (Vol. 9, pp. 399-432). Elsevier.

57. Fairbridge, R. W., (1967). *The Encyclopedia of Atmospheric Sciences and Astrogeology* (Vol. 1, p. 1200). New York: Reinhold.

58. Fairbridge, R. W., (1968). *The Encyclopedia of Geomorphology* (Vol. 1, p. 1295). New York: Van Nostrand Reinhold.

59. Fairbridge, R. W., (1971). Upper Ordovician glaciation in Northwest Africa? *Geol. Soc. Am. Bull., 82*, 269–274.

60. Fairbridge, R. W., (1973). Glaciation and plate migration. In: Tarling, D. H., & Runcorn, K., (eds.), *Implications of Continental Drift to the Earth Sciences* (Vol. 1, pp. 503–515). London: Academic Press.

61. Fairbridge, R. W., (1975). *The Encyclopedia of World Regional Geology* (Vol. 1, p. 704). Part 1: Western Hemisphere. Stroudsburg, Dowden: Hutchinson & Ross.

62. Fairbridge, R. W., (1977). Rates of sea ice erosion of Quaternary littoral platforms. *Studia Geol. Polonica, 52*, 135–141.

63. Fairbridge, R. W., (1982). The fracturing of Gondwanaland. In: Scrutten, R. A., & Talwani, M., (eds.), *The Ocean Floor* (Vol. 1, pp. 229–235). Chichester, NY: Wiley.

64. Fairbridge, R. W., (1990). Solar and lunar cycles embedded in the El Niño periodicities. *Cycles, 41*, 66–72.

65. Ferretti, D. F., et al., (2005). Unexpected changes to the global methane budget over the past (2000) years. *Science, 309*, 1714–1717.

66. Fischer, G., & Wefer, G., (1999). *Use of Proxies in Paleoceanography: Examples from the South Atlantic* (Vol. 1, p. 735). Springer, Berlin.

67. Fletcher, C. H. III, et al., (1993). Emergence of the Varanger Peninsula, Arctic Norway, and climate changes since deglaciation. *The Holocene, 3*(2), 116–127.

68. Flint, R. F., (1971). *Glacial and Quaternary Geology* (Vol. 1, p. 892). New York: Wiley.

69. Flohn, H., & Fantechi, R., (1984). *The Climate of Europe: Past, Present, and Future* (Vol. 1, pp. 1–24). Dordrecht, The Netherlands: Reidel Publishing Co.

70. Flückiger, J., et al., (1999). Variations in atmospheric N_2O concentration during abrupt climatic changes. *Science, 285*(5425), 227–230.

71. Flückiger, J., et al., (2002). High resolution Holocene N_2O ice core record and its relationship with CH_4 and CO_2. *Global Biogeochem. Cycles*, *16*, 1–20. doi: 10.1029/(2001)GB001417.

72. Frakes, L. A., (1979). *Climates Throughout Geologic Time* (Vol. 1, p. 379). Amsterdam: Elsevier.

73. Francois, L. M., Delire, C., Warnant, P., & Munhoven, G., (1998). Modelling the glacial-interglacial changes in the continental biosphere. *Global Planet. Change*, *17*, 37–52.

74. Francois, L. M., et al., (1999). Carbon stocks and isotopic budgets of the terrestrial biosphere at mid-Holocene and last glacial maximum times. *Chem. Geol.*, *159*, 163–189.

75. Freeman, K. H., & Hayes, J. M., (1992). Fractionation of carbon isotopes by phytoplankton and estimates of ancient CO2 levels. *Global Biogeochem. Cycles*, *6*, 185–198.

76. Genty, D., & Quinif, Y., (1996). Annually laminated sequences in the internal structure of some Belgian stalagmites; importance for paleoclimatology. *Journal of Sedimentary Research*, *66*(1), 275–288.

77. Gohau, G., (1990). *A History of Geology* (Vol. 1, p. 259). New Brunswick, NJ: Rutgers University Press.

78. Grove, J. M., (1988). *The Little Ice Age* (Vol. 1, p. 498). London: Methuen.

79. Guo, Z. T., et al., (2004). Late Miocene-Pliocene development of Asian aridification as recorded in the Red-Earth formation in northern China. *Global Planet. Change*, *41*(3–4), 135–145.

80. Hassan, F. A., (1981). Historical Nile floods and their implications for climate change. *Science, 212*, 1142–1144.

81. Heer, O., (1865/1876). *The Primeval World of Switzerland, London* (Vol. 2, pp. 1–24).

82. Herschy, R. W., & Fairbridge, R. W., (1998). *Encyclopedia of Hydrology and Water Resources* (Vol. 1, p. 803). Dordrecht: Kluwer Academic Publishers.

83. Hoyle, F., & Littleton, R. A., (1919). The effect of interstellar matter on climatic variation. *Proceedings of the Cambridge Philosophical Society, 35*, 405–418.

84. Huntington, E., & Visher, S. S., (1922). *Climatic Changes, Their Nature and Causes* (Vol. 1, p. 329). New Haven, CT: Yale University Press.

85. Imbrie, J., & Imbrie, K. P., (1979/1986). *Ice Ages: Solving the Mystery* (Vol. 1, p. 224). Cambridge, MA: Harvard University Press.

86. Isacks, B. L., Oliver, J., & Sykes, L. R., (1968). Seismology and the new global tectonics. *J. Geophys. Res., 73*, 5855–5899.

87. Jelbring, H., (1998). *Wind Controlled Climate* (Vol. 1, p. 111). PhD Thesis, Akademitryk AB, Sweden: Stockholm University.

88. Joos, F., et al., (2004). Transient simulations of Holocene atmospheric carbon dioxide and terrestrial carbon since the Last Glacial Maximum. *Global Biogeochem. Cycles*, *18*, 1–30. doi: 10.1029/(2003)GB002156.

89. Kageyama, M., et al., (2006). Last glacial maximum temperatures over the North Atlantic, Europe, and western Siberia: A comparison between PMIP models, MARGO sea-surface temperatures and pollen-base reconstructions. *Quat. Sci. Rev.*, *25*, 2082–2102.

90. Kaplan, J. O., et al., (2003). Climate change and Arctic ecosystems: 2. Modeling, paleodata-model comparisons, and future projections. *J. Geophys. Res.*, *108*, 1–27. doi: 10.1029/(2002)JD002559.

91. Kaplan, J. O., Prentice, I. C., Knorr, W., & Valdes, P. J., (2002). Modeling the dynamics of terrestrial carbon storage since the last glacial maximum. *Geophys. Res. Lett.*, *29*, 1–29. doi: 10.1029/(2002) GL015230.

92. Keeling, C. D., & Whorf, T. P., (2005). Atmospheric CO_2 records from sites in the SiO air sampling network. In: *Trends: A Compendium of Data on Global Change* (Vol. 1, pp. 1–23). Carbon dioxide information analysis center, Oak Ridge National Laboratory, U.S. Department of Energy, Oak Ridge, TN.

93. Kennett, J. P., & Stott, L. D., (1991). Abrupt deep-sea warming, paleoceanographic changes and benthic extinctions at the end of the Paleocene. *Nature*, *353*, 225–229.

94. Kennett, J. P., (1982). In: Prentice-Hall & Kerner-Marilaun, F., (eds.), *Marine Geology* (Vol. 1, pp. 1–24). Englewood Cliffs, NJ: "Paläoklimatologie." Berlin: Gebr. Borntraeger.

95. Kerr, R. A., (2006). A worrying trend of less ice, higher seas. *Science, 311*, 1698–1701.

96. Koch, P. L., Zachos, J. C., & Gingerich, P. D., (1992). Correlation between isotope records in marine and continental carbon reservoirs near the Paleocene Eocene boundary. *Nature*, *358*(6384), 319–322.
97. Koerner, R. M., (1989). Ice core evidence for extensive melting of the Greenland Ice-sheet in the last interglacial. *Science*, *244*(4907), 964–968.
98. Kukla, G. J., (1972). Insolation and glacials. *Boreas, 1*, 63–96.
99. Lamb, H. H., & Woodroffe, A., (1970). Atmospheric circulation during the last ice age. *Quaternary Res., 1*, 29–58.
100. Lamb, H. H., (1977). *Climate History and the Future* (Vol. 1, p. 835). London: Methuen (and Princeton University Press).
101. Lewis, R. S., & Anders, E., (1985). Cretaceous extinctions: Evidence for wildfires and search for meteoritic material. *Science, 230*, 167–170.
102. Lyell, C., (1830–1833). *Principles of Geology* (Vol. 3, pp. 1–23). London: John Murray. [facsimile reprint, New York: Johnson Reprint Corp., 1969, introduction by M. J. S. Rudwick].
103. Maclaren, C., (1842). The glacial theory of Professor Agassiz of Neuchatel. *Am. J. Sci., 42*, 346–365.
104. Mahowald, N., et al., (1999). Dust sources and deposition during the last glacial maximum and current climate: A comparison of model results with paleodata from ice cores and marine sediments. *J. Geophys. Res.*, *104*, 15859–15916.
105. Maley, J., (1997). Middle to late Holocene changes in tropical Africa and other continents: Paleomonsoon and sea surface temperature variations. In: Dalfes, H. N., Kukla, G., & Weiss, H., (eds.), *Third Millennium BC Climate Change and Old-World Collapses* (Vol. 1, pp. 611–640). Berlin: Springer-Verlag.
106. Mangerud, J., Astakhov, V., & Svendsen, J. I., (2002). The extent of the Barents-Kara ice sheet during the last glacial maximum. *Quat. Sci. Rev.*, *21*, 111–119.
107. Marvin, U. B., (1973). *Continental Drift, the Evolution of a Concept* (Vol. 1, p. 205). Washington, D.C.: Smithsonian Institute Press.
108. Masson-Delmotte, V., et al., (2006). Past and future polar amplification of climate change: Climate model intercomparisons and ice-core constraints. *Clim. Dyn.*, *26*(5), 513–529.

109. McElwain, J. C., & Chaloner, W. G., (1995). Stomatal density and index of fossil plants track atmospheric carbon dioxide in the Paleozoic. *Ann. Bot. (London)*, *76*, 389–395.

110. McGovern, T. H., (2000). The demise of Norse Greenland. In: Fitzhugh, W. W., & Ward, E. I., (eds.), *Vikings: The North Atlantic Saga* (Vol. 1, pp. 327–393). Washington and London: Smithsonian Institution Press and National Museum of Natural History.

111. McManus, J. F., et al., (2002). Thermohaline circulation and prolonged interglacial warmth in the North Atlantic. *Quat. Res.*, *58*, 17–21.

112. Milankovitch, M., (1941). *Kanon der erdbestrahlung und seine Anwendung auf das Eiszeitenproblem* (p. 484, 633). Roy. Serb. Acad., Special Publication 133 (Engl. Transl. 1969 as *"Canon of Insolation and the Ice-Age Problem,"* by Israel Progr. Sci. Transl., U.S. Dept. Commerce, Washington).

113. Moberg, A., (1995). Multi-taper spectral analysis of the Stockholm air temperature record: A significant 22.8 yr cyclic pulse observed. *J. Coast. Res.*, *17*, 39–44.

114. Monnin, E., et al., (2001). Atmospheric CO_2 concentrations over the last glacial termination. *Science*, *291*(5501), 112–114.

115. Monnin, E., et al., (2004). Evidence for substantial accumulation rate variability in Antarctica during the Holocene, through synchronization of CO_2 in the Taylor Dome, Dome C and DML ice cores. *Earth Planet. Sci. Lett.*, *224*(1–2), 45–54.

116. Moran, K., et al., (2006). The Cenozoic paleoenvironment of the Arctic Ocean. *Nature*, *441*, 601–605.

117. Mörner, N. A., (1984). Low sea levels, droughts, and mammalian extinctions. In: Berggren, W. A., & Van Couvering, J. A., (eds.), *Catastrophes and Earth History* (pp. 387–394). Princeton, NJ: Princeton University Press.

118. Mudelsee, M., (2001). The phase relations among atmospheric CO_2 content, temperature, and global ice volume over the past 420 ka. *Quat. Sci. Rev.*, *20*, 583–589.

119. Muhs, D. R., Simmons, K. R., & Steinke, B., (2002). Timing and warmth of the last interglacial period: New U-series evidence from Hawaii and Bermuda and a new fossil compilation for North America. *Quat. Sci. Rev.*, *21*, 1355–1383.

120. Nansen, F., (1922). *The Strandflat and Isostasy* (No. 11, pp. 1–32). I Kommission hos J. Dybwad.

121. Nicholson, S., & Flohn, H., (1980). African environmental and climatic changes and the general circulation in late Pleistocene and Holocene. *Climatic change, 27*, 20–34.

122. Nölke, F., (1919). *The Problem of the Evolution of Our Planetary System: A Critical Study* (Vol. 1, pp. 1–23). Jumper.

123. Officer, C., & Page, J., (1996). *The Great Dinosaur Extinction Controversy* (Vol. 1, p. 209). Reading, MA: Addison & Wesley.

124. Oliver, J. E., & Fairbridge, R. W., (1987). *The Encyclopedia of Climatology* (Vol. 1, p. 986). New York: Van Nostrand Reinhold.

125. Peixoto, J. P., & Oort, A. H. (1992). *Physics of Climate* (Vol. 1, p. 512). New York: American Institute of Physics.

126. Petit, J. R. et al., (1999). Climate and atmospheric history of the past 420,000 years from the Vostok ice core, Antarctica. *Nature, 399*, 429–436.

127. Pettersson, O., (1912). The connection between hydrographical and meteorological phenomena. *Quarterly J. Roy. Meteor. Soc., 38*, 173–191.

128. Pettersson, O., (1930). The tidal force: Geogr. *Annaler, 12*, 261–322.

129. Pitman, W. C. III, & Heirtzler, J. R., (1966). Magnetic anomalies over the Pacific-Antarctic Ridge. *Science, 154*, 1164–71.

130. Pitman, W. C., (1978). Relationship between eustasy and stratigraphic sequences of passive margins. *Geol. Soc. Am. Bull., 89*, 1389–1403.

131. Poag, C. W., (1997). The Chesapeake Bay bolide impact: A convulsive event in Atlantic coastal plain evolution. *Sediment. Geol., 108*, 45–90.

132. Rampino, M. R., & Haggerty, B. M., (1996). "The 'Shiva Hypothesis': Impacts, mass extinctions and the galaxy." *Earth, Moon, and Planets, 72*, 441–460.

133. Rampino, M. R., (1998). The galactic theory of mass extinctions: An update. *Celestial Mech. Dyn. Astron. 69*, 49–58.

134. Raymo, M. E., & Ruddiman, W. F., (1992). Tectonic forcing of late Cenozoic climate. *Nature, 359*, 117–122.

135. Raymo, M. E., (1994). The initiation of northern hemisphere glaciation. *Annu. Rev. Earth Planet. Sci., 22*, 353–383.

136. Rind, D., (2002). The sun's role in climate variations. *Science, 296*, 673–677.

137. Rognon, P., & Williams, M. A. J., (1977). Late quaternary climatic changes in Australia and North Africa: A preliminary interpretation. *Paleogeography, Paleoclimatology, Paleoecology*, *21*(4), 285–327.

138. Ruddiman, W. F., (2005). *Plows, Plagues, and Petroleum: How Humans Took Control of Climate* (Vol. 1, p. 272). Princeton, NJ: Princeton University Press.

139. Ryan, W. B.F., & Pitman, W. C., (1999). *Noah's Flood: The New Scientific Discoveries About the Event That Changed History* (Vol. 1, p. 320). New York, NY: Simon & Schuster.

140. Sayles, R. W., (1931). Bermuda during the ice age. *Proceedings of the American Academy of Arts & Sciences, 66*, 382–467.

141. Schove, D. J., (1983). *Sunspot Cycles (Benchmark Papers in Geology)* (Vol. 68, p. 392). Stroudsburg: Hutchinson Ross Publ. Co.

142. Schramm, M., (1963). Aristotelianism: Basis and obstacle to scientific progress in the middle ages. *History of Science, 2*, 91–113.

143. Schwarzbach, M., (1974). *Das Klimate der Vorzeit* (Vol. 3, pp. 1–32). Stuttgart: F. Enke Verlag, (English edition, 1963). New York: Van Nostrand.

144. Shindell, D. T., Schmidt, G. A., Mann, M. E., Rind, D., & Waple, A., (2001). Solar forcing of regional climate change during the maunder minimum. *Science, 294*, 2149–2152.

145. Shirley, J. H., & Fairbridge, R. W., (1997). *Encyclopedia of Planetary Sciences* (Vol. 1, p. 990). London: Chapman & Hall.

146. Sigurdsson, H., Houghton, B. F., McNutt, S. R., Rymer, H., & Stix, J., (2000). *Encyclopedia of Volcanoes* (Vol. 1, p. 1417). San Diego, CA: Academic Press.

147. Siscoe, G. L., (1980). Evidence in the auroral record for secular solar variability. *Rev. Geophys. Space Phys., 18*, 647–658.

148. Stommel, H., (1958). *The Gulf Stream* (Vol. 1, p. 202). Berkeley, CA: University of California Press.

149. Street, F. A., & Grove, A. T., (1976). Environmental and climatic implications of late quaternary lake-level fluctuations in Africa. *Nature, 281*, 380–385.

150. Street, F. A., & Grove, A. T., (1979). Global maps of lake-level fluctuations for 30,000 years BP. *Quaternary Res., 12*, 83–118.

151. Stuiver, M., & Braziunas, T. F., (1989). Atmospheric ^{14}C and century-scale solar oscillations. *Nature, 388*, 405–408.

152. Suess, E., (1909). *The Face of the Earth: The Face of the Earth* (Vol. 4, pp. 1–32). Clarendon Press.

153. Thunell, R. C., Williams, D. F., & Kennett, J. P., (1977). Late quaternary paleoclimatology, stratigraphy and sapropel history in eastern Mediterranean deep-sea sediments. *Marine Micropaleontology, 2*, 371–388.

154. Valentine, J. W., & Moores, E. M., (1970). Plate-tectonic regulation of faunal diversity and sea level: A model. *Nature, 228*, 659–669.

155. Visher, G. S., (1986). A history of geological thought. *Earth Sciences History, 5/2*, 137–143.

156. Von Richthofen, F., (1877–1885). *Results of Own Trips and Studies Based on Them* (Vol. 4, pp. 1–34). Berlin.

157. Wegener, A., (1929). *The Formation of the Continents and Oceans* (Vol. 4, p. 346). Vieweg & Sohn Akt. Ges, Braunschweig, New York: Dover Publications.

158. Wendorf, F., & Schild, R., (1976). *Prehistory of the Nile Valley* (Vol. 1, p. 404). New York: Academic Press.

159. Williams, M. A. J., & Faure, H., (1980). *The Sahara and the Nile* (Vol. 1, p. 607). Rotterdam: A.A. Balkema.

160. Wood, F. J., (2001). *Tidal Dynamics* (3rd edn., Vol. 2, pp. 1–29). West Palm Beach: Coastal Education and Research Foundation.

161. Zenkovitch, A., (1966). *The Encyclopedia of Oceanography* (Vol. 1, p. 146). New York: Reinhold.

162. Zeuner, F. E., (1959). *The Pleistocene Period* (Vol. 2, p. 447). Hutchinson & Co. London: Royal Society.

Chapter 3

Basics of Paleobotany

CONTENTS

3.1. Introduction 96
3.2. Nearest-Living-Relative Technique 97
3.3. Plant Structure Proxies 99
3.4. Leaves 100
3.5. Wood 103
3.6. Chemical Signatures 105
References 107

3.1. INTRODUCTION

Because plants are sessile organisms, they are perfect as recorders of, or representations for, terrestrial climate. They must adapt or perish as the environment changes around them. Many features of plant growth have been restricted through adaption to precise climate regimes over evolutionary time (such as, Loucks and Parkhurst, 1972; Creber and Chaloner, 1990; Mosbrugger and Uhl, 1999). Plant fossils could be utilized as a record of local paleoclimate because of their sessile nature and the tendency for several plant parts to be conserved very close to where they grew (Figure 3.1) (Spicer, 1989; Greenwood and Wing, 1995).

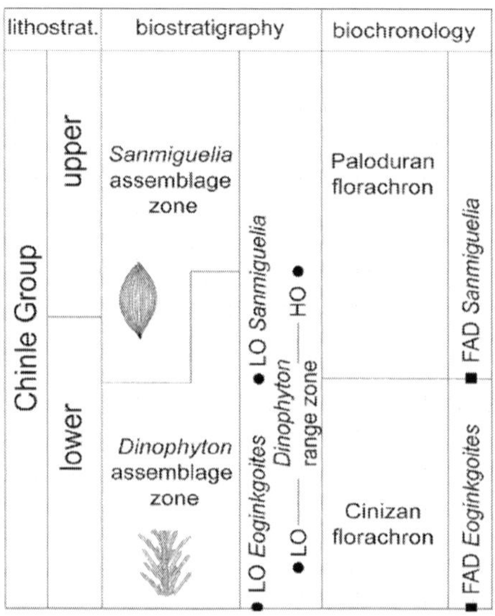

Figure 3.1. With the Chinle group as an example, some basic ideas of paleobotanical biostratigraphy and biochronology.

Source: https://www.researchgate.net/figure/Some-basic-concepts-of-paleobotanical-biostratigraphy-and-biochronology-with-the-Chinle_fig2_280880438.

Paleoclimate from fossil plants is usually determined using three methods. The first method is known as the nearest living-relative method. The nearest-living-relative method likens fossil plants to the existing plants to which they are most closely linked, assuming that both have similar climatic tolerances over geologic time. The second method makes use of plant features as climate proxies. These methods use various features of wood and leaves as proxies to find the climate in which these plant parts grew, like stomatal index, tree rings, and leaf morphology. The last method involves determining the plant's climatic surroundings by detecting isotopic ratios in plant tissue.

3.2. NEAREST-LIVING-RELATIVE TECHNIQUE

In paleoclimate reconstructions, the nearest-living-relative method is commonly employed. This taxon-based technique compares ancient plants to the closely related current plants, generally established through morphological similarities among a living taxon and a fossil. The essential expectations are that both modern plants and fossils have comparable climatic tolerances and that the relationships and identifications assumed among extant taxa and fossils are valid. According to Dilcher (1974, 2000), 60 to 75% of taxa assigned to Cretaceous fossil angiosperms and Paleogene are inaccurate. This method has the advantage of working with a wide range of plant organs, including wood, pollen, leaves, seeds, and fruits. Multiple plant organs can be employed to produce a stronger paleoclimate determination for specific plant assemblages. Because the affinity of taxonomic is often challenging to detect with leaf material, assemblages including seeds and fruits or mixtures of plant parts produce more precise outcomes than those with only leaves. When a high number of taxa are employed, the taxonomic identifications are at the species level, and the taxonomic identifications are correct, the best results are produced. This method implies that the fossil and current relatives' ecological tolerances are comparable and have not altered through time. Recently a fossil is, the more possible the ecological tolerances are comparable. As a result, the Neogene and Holocene climate reconstructions are more expected to be correct than the Mesozoic and Paleogene comparisons. Paleozoic plants are too different from modern plants to employ this method.

Since ecological tolerances of types are frequently more established than those of larger taxonomic groups, comparisons at the species level are the most reliable. However, in several cases, especially when utilizing wood or pollen, the fossil can only be identified to the genus or family level. Tolerances are comparable across genera and families in several cases. Gingers and palms, for instance, have been believed to grow and reproduce in tropical and subtropical environments exclusively, even though palms frequently reproduce and grow in warm temperate zones of southeastern South America and North America. Like, Acer, and Quercus, some genera have extensive ecological tolerances, and most plant families have a comprehensive environmental range (e.g., Fabaceae). This difficulty is mitigated to some extent by reconstructing paleoclimate using many taxa, which minimizes the possibility of a single broad-ranging taxon influencing the results.

Both calculate paleoclimate and absolute values and determine climatic or vegetation zones, and the nearest-living-relative technique is applied. The former method needs knowledge of several diverse modern species' environmental tolerances, whereas the latter depends on knowing which taxon combinations are frequently found together in current climate or vegetation zones. Individual modern taxonomic relatives of fossil organisms can determine absolute climatic values, resulting in a range of temperatures or precipitation levels. Combining several modern climate ranges, on the other hand, should yield a more exact estimation for the fossil accumulation, along with the subsequent climatic estimate being the range in which all current taxa align (Figure 3.2). When multiple taxonomic ranges are combined, the risk of a single problematic taxon skewing the final climate range is reduced. The coexistence technique (Mosbrugger and Utescher, 1997) is the application of this strategy that employs a database of modern ecological tolerances of seeds, fruits, and leaves.

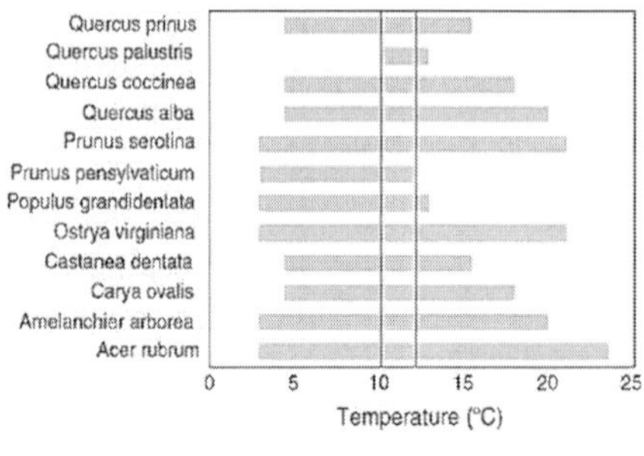

Figure 3.2. From southern Pennsylvania a hypothetical list of modern relatives based on flora.

Source: https://link.springer.com/referenceworkentry/10.1007%2F978-1-4020-4411-3_165.

Note: If these current taxa had signified the current relatives of fossil flora, the projected mean annual temperature of the fossil site would be between around 10°C and 12°C, the range of overlap among all taxa present.

The nearest-living-relative approach will be used to fossil assemblages throughout Europe. This method has yielded temperature estimations with ranges that is very small in the range of 1–4°C for Tertiary and younger-aged fossil accumulations.

The approach of determining climatic and vegetation zones employing the nearest-living-relative method is more popular than discovering absolute climate values, as palynologists favor it. Comparable taxonomic unions in both fossil and modern assemblages are used in this application of the approach. Using a large number of species, as with the coexistence technique, is preferred when defining a paleoclimatic or paleovegetation zone. This is particularly true when employing pollen grains, that are sometimes only identified at the level of genus and have excellent discrete ecological tolerances but a lower assemblage tolerance. Nevertheless, independent of the other taxa involved, the existence of a limited number of highly suggestive taxa can sometimes compel the paleoclimatic or paleovegetation zone (Martin, 1997). Atkinson and Sinka (1999) employed factor analysis to identify overlying climate zones in contemporary plant assemblies and then map unidentified fossil ranges on a "climate space." This method is a more objective way of establishing climatic zones in fossil assemblies, as it eliminates any partiality that one taxon might have in the entire zonation.

3.3. PLANT STRUCTURE PROXIES

Even in floras that are geographically and compositionally isolated, equivalent combinations of wood or leaf features are common in floras that dwell in comparable climates. As a result, precise plant traits arise as developing flora properties in a particular climate regime. The absence or presence of these features in a fossil floral assembly can be used to determine the assemblage site's paleoclimate. The current correlations between climate and various plant traits are used to interpret climate. Distinct from the nearest-living-relative technique, wood, and Foliar physiognomy procedures do not need precise or correct taxonomic identification. However, they require knowledge of the number of taxa present and some characteristics of their anatomical or morphological traits. This can be accomplished by employing a parataxonomical or morphotype-based method (Woodcock and Ignas, 1994; Davies-Vollum, 1997), which interprets groupings of morphologically similar plant fossils as species groups. Leaf venation density and stomatal indices approaches require precise taxonomic identification because they are species-dependent proxies.

3.4. LEAVES

Stomatal index, stomatal density or leaf physiognomy, and leaf venation density are all common proxies used in paleoclimatic reconstructions of angiosperm leaves. Seasonal and annual temperature and precipitation are frequently determined using leaf physiognomy. Temperature, moisture availability, and PaleoCO_2 levels are all determined using stomatal index and stomatal density. The density of the leaf venation is employed to calculate transpiration stress.

Leaf physiognomic approaches are based on several qualitative correlations between climate and leave measured by employing floral data and modern climatic changes from various climates and regions. In addition, analytical equations have been developed to identify climate among fossil leaves by calculating the quantities of particular leaf features found in specific modern climates. The most popular equations correlative leaf morphological characteristics to precipitation and mean annual temperature, although others have been employed to measure other climatic factors like increasing season precipitation and warm and cold month mean temperatures.

Simple linear regression has been proved to be a comparatively accurate and direct method of forecasting precipitation and yearly mean temperature. The relationship between the state of the leaf margin (smooth or toothed) and the mean annual temperature is strong (Figure 3.3), resulting in an equally accurate and precise prediction equation. As a result, this has been the most investigated and perfected relationship. The derived mean annual temperature estimations have a minimal typical error of 2°C (Wilf, 1997). In rain equations connecting leaf area to mean annual rain has shown the most potential and are being revised by many scientists (such as, Jacobs, 2002). Though, because the estimations contain significant standard errors, most researchers have advised care depending on the resulting mean annual precipitation predictions.

Because the geomorphology of most leaf features is assumed to be the result of the intricate interface of several climate elements, numerous authors have begun to use a multi-character method to climate forecast from leaf morphology. Multiple regression equations with 2 to 10 leaf features have been shown to improve climate prediction (e.g., Wing and Greenwood, 1993; Gregory-Wodzicki, 2000; Jacobs, 2002) and are often more accurate than simple linear regression equations for paleoclimate forecast. CLAMP is a vast, global database of floral data and modern climate created by Wolfe (1993) a multivariate statistical method is used for canonical correspondence analysis for paleoclimate prediction. His catalog predicts various climate factors, such as growing season temperature and precipitation, mean annual temperature, enthalpy, and temperature range, using 31 leaf features. To develop paleoclimate prognostic equations, nearest neighbor analysis analyses the CLAMP database sites closest to the unidentified site in multivariate space (Stranks and England, 1997). By limiting the predictor set to similar floras, this method has enhanced multivariate prediction even further, frequently resulting in minor estimate error.

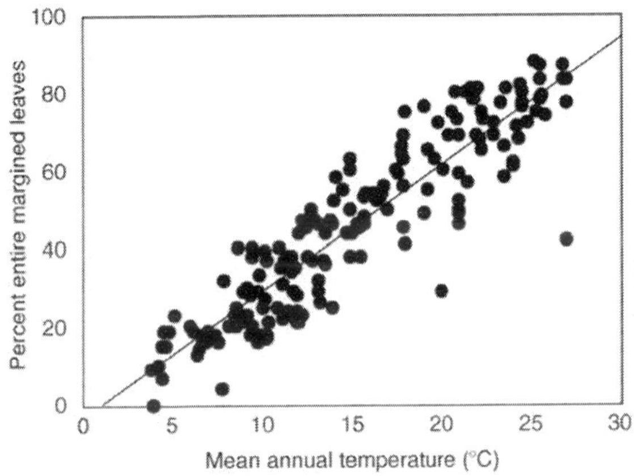

Figure 3.3. Percentage of whole margined leaves in a flora versus mean annual temperature for 177 sampled floras global (Wolfe, 1993; Gregory-Wodzicki, 2000).

Source: https://pubs.er.usgs.gov/publication/b2040.

Predictions based on leaf physiognomy are reasonably accurate, although certain assumptions must be made. The initial is that the association between climate and leaf morphology has remained stable across time. The 2nd point is that they are minor while taphonomic biases occur (Dilcher and Roth, 1978; Greenwood, 1992). The 3rd point is that there are not many regional variances because of climate or taxonomical history. Numerous writers have indicated that it is critical to employ predictor data sets comparable to those found in the fossil assemblage in terms of geography, taxonomy, and climate. This constraint should mitigate the issues that the third assumption above might cause. To reduce paleoclimate, estimate errors, leaf physiognomy approaches require many fossil taxa (20). Only the leaf features of woody angiosperms have been proven to be linked to climate, limiting the method's application to the Cenozoic and Cretaceous periods.

To measure moisture availability, temperature, or atmospheric CO_2 concentrations, stomatal density and stomatal index can be utilized. Guard cells surround stomata pores, regulating the stomatal opening's dimensions and hence gas exchange and water loss from the leaf to the atmosphere. When the leaf acts as a biosensor of ambient CO_2, the number of stomata that distinguish is determined at the moment of leaf unfolding. Stomatal index is more commonly used than stomatal density since it removes leaf size on density counts. Among other species, Metasequoia, Ginkgo, and Quercus have been employed to standardize CO_2-stomatal index curves from which PaleoCO_2 values have been estimated. All species exhibit a negative association between ambient carbon content and stomatal index and stomatal density. These organisms have a broad temporal range (both Metasequoia and Ginkgo are referred to as "living fossils") and may follow changes in atmospheric carbon over an extended time. Stomatal density and stomatal index, like other approaches, provide benefits but should be used with care. The stomatal index responds non-linearly when CO_2 levels rise. As a result, issues develop when determining PaleoCO_2 during periods when atmospheric CO_2 concentrations were higher than they are now (Kurschner et al., 1998; Beerling and Royer, 2002). Even at the same CO_2 level, the stomatal index drops as the temperature rises (Beerling and Chaloner, 1993). If the temperature does not remain constant across the time being studied, this can impact CO_2 estimates in the atmosphere. Temperature can, however, be determined in conjunction with other leaf physiognomy approaches, reducing any interpretation issues. Stomatal density appears to be related to precipitation amount, at least qualitatively, notably in shade leaves (Sun et al., 2003). On the other hand, the stomatal index does not appear to imitate this relationship and is a superior indicator of the concentration of atmospheric CO_2. The fact that stomatal index and stomatal density can be used to determine CO_2 from the Paleozoic to the Recent Period is a considerable advantage. Leaf venation density is a technique for estimating paleomoisture parameters that have just recently been developed. The density of leaf venation variations in response to the quantity of leaf transpiration, which might be caused by a lack of water or a temperature change. Because climate conditions influence leaf venation density during leaf development in a single growing season, this approach can detect small-scale alterations that differ by season. However, this method still has various flaws, including density variances based on leaf size, measuring density, and disparities between sun and shade leave. However, this approach might be employed from the Paleozoic to the Recent if these issues are resolved.

3.5. WOOD

Charcoal, tree rings, and wood anatomy are only a few of the features of wood that can be utilized to estimate paleoclimate. In addition, wood is frequently found in ancient territories that are not conducive to preserving leaves. As a result, employing wood as a climate proxy expands the range of situations under which it is possible to recreate paleoclimate from plants.

The size of vessel components, arrangement, and presence, fiber cells, axial parenchyma, and ray cells are all used to recreate paleoclimates from wood anatomy (Wiemann et al., 1999; Ignas and Woodcock, 1994). Relationships detected among diameter with moisture and vessel density, temperature-dependent vessel diameter, and the existence of ring porosity as a pointer of seasonality have all been used to establish the relationship between wood anatomy and climate. Numerical equations have been developed, especially by Wiemann et al. (1998, 2001), which produce numerical climate estimations based on these same properties. These equations should foresee temperature to inside 5°C, yet equations employing different predictive types from the same site can vary by nearly 6°C. The temperature has a good correlation with anatomical characteristics of wood, whereas precipitation has a weak correlation (Wiemann et al., 1998, 2001). To estimate climate, these equations require many taxa (~25), which reduces inaccuracy. Terral and Mengal (1999) demonstrated that if a species exhibits variation in anatomical features concerning climate, it is possible to utilize that species to estimate climate across time. Olive wood was used to determine Holocene paleoclimate in their research.

Climate estimates based on wood anatomy, both quantitative and qualitative, function reasonably well to identify hot from more seasonal or temperate locales. Climate has a more significant impact on the cellular structure of wood grownup in seasonal or temperate climates than on wood cultivated in tropical climates. As a result, utilizing wood to reconstruct climate permits for extensive categorical generalizations about paleoclimate and can differentiate effectively among global sites. The taxonomic intersection among sites can result in temperature changes that are not represented in the wood assembly in smaller areas (e.g., eastern North America) (Kowalski, personal observation). Anatomical features utilized in climate reconstruction, like the absence or presence of curved thickening, do not differ considerably between species in some places.

The wood structure includes the same cautions as leaf approaches when determining paleoclimate, like the hypothesis that the relationship among anatomy and climate has remained stable across time. Wood structure methods may theoretically be utilized back to the Paleozoic. Few wood features that are employed to guess climate are, however, derived. Because these wood features evolved with the emergence of angiosperms during the late Mesozoic epoch (Francis and Poole, 2002), they cannot be employed to estimate Paleozoic and middle and early Mesozoic paleoclimates.

Because each ring records temperature and moisture characteristics for one growing season, dendrochronology, also known as tree ring analysis, can be employed to control climate over numerous seasons.

The information that results describes a lengthy sequence of yearly climate and climate changes with a level of accuracy that few other records can approach. The majority of data on how tree ring features respond to climate comes from conifer research (Francis and Poole, 2002).

However, angiosperms are periodically studied too. In the southwestern United States, a constant record of yearly tree rings dating back over 9,000 years has been constructed. Changes in local climatic moisture and temperature are frequently linked to the earliest human inhabitances and migrations.

The existence or lack of tree rings indicates seasonality, which is generally linked to temperature (Poole and Francis, 2002). Because some humid tree species have rings and can be muddled with temperate species, ring presence as a sign of temperature seasonality is best assessed from many taxa.

Seasonal leaf loss or seasonality of precipitation is the most plausible cause of tropical tree rings (Woodcock and Ignas, 1994). Wide rings result from warmer, surfactant conditions or more CO_2 (Beerling, 1999), while narrow rings result from drier, cooler conditions (Poole and Francis, 2002). False rings can occasionally appear, indicating a pause in the increasing season because of insect damage, drought, fire, or cold. Instead of determining a single flora's absolute temperature or humidity, ring widths are finest employed for wide floral comparisons.

The fluctuation of tree rings from year to year is measured by mean sensitivity, which indicates the consistency of the climate regime (Poole and Francis, 2002). Tree ring studies can be executed on floras from the new through Paleozoic periods; however, most applications focus on climate over the previous 10,000 years.

Lastly, whether PaleoO_2 levels are inside the comparatively restricted range relate to charcoal formation can be determined by the existence of charcoal in an assemblage. Only a tiny amount of oxygen is required for charcoal to form (Chaloner and McElwain, 1997). Because plant pieces are regularly charcoalified and found in Devoniantothe Recent strata, additional research is needed in this area. Mesofossils are small remnants employed in floristic investigations to ascribe climate affinities to the closest extant relative.

3.6. CHEMICAL SIGNATURES

During their growth, plants absorb oxygen, carbon, and hydrogen isotopes from CO_2 in the atmosphere and water. The plant incorporates these isotopes into its structure. Thus, they serve as a record of the plant's environment and climate. Through comparing the values of tree rings over the tree's life, these records can illustrate long-term fine-scale fluctuations or show specific season values through looking at individual tree rings or leaves. $d^{13}C$ measurements have been employed to link CO_2 records in tree rings along with ice core data to identify the precise ages of the trees being calculated, in addition to estimating paleoclimate.

The most frequent isotopic ratio employed to control paleoclimate from plants is ^{13}C, which is absorbed into the plant during growth from CO_2 in the environment. ^{13}C ratios in the cellulose of wood are measured and indicate a negative connection with temperature and relative humidity (Edwards et al., 2000). Sun et al. (2003) found a negative connection between ^{13}C ratios and humidity or precipitation using ^{13}C readings from the leaf cuticle (Figure 3.4).

In addition to d^2H, $d^{18}O$ and ^{13}C have been detected in swift water, seasonal moisture (precipitation and humidity), and precipitation amount utilizing wood cellulose and wood nitrite (Buhay and Edwards, 1995; Edwards et al., 2000). Furthermore, d^2H links with seasonal temperature and mean annual; however, this association is best evaluated in places where the water supply is isotopically consistent (Buhay and Edwards, 1995). Climate estimates based on simply one isotope may be misleading due to the complicated relationship between precipitation and temperature. The most excellent paleoclimatic information from plants appears to be obtained by combining $d^{18}O$, d^2H, and $d^{13}C$ observations (Edwards et al., 2000).

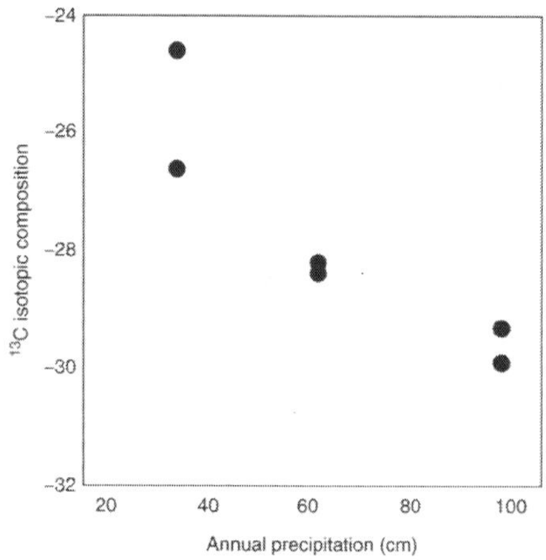

Figure 3.4. For three sites in China, the ^{13}C isotopic composition of current Ginkgo leaves was compared to yearly precipitation. (Sun et al., 2003).

Source: https://www.pnas.org/content/100/12/7141.

REFERENCES

1. Beerling, D. J., & Chaloner, W. G., (1993). The impact of atmospheric CO_2 and temperature change on stomatal density: Observations from *Quercus robur* Lammas leaves. *Ann. Bot., 71*, 231–235.
2. Beerling, D. J., & Royer, D. L., (2002). Reading a CO_2 signal from fossil stomata. *New Phytol., 153*, 387–397.
3. Beerling, D. J., (1999). Atmospheric carbon dioxide, past climates, and the plant fossil record. *Bot. J. Scotl., 51*, 49–68.
4. Buhay, W. M., & Edwards, T. W.D., (1995). Climate in southwestern Ontario, Canada between AD 1610 and 1885 inferred from oxygen and hydrogen isotopic measurements of wood cellulose from trees in different hydrologic settings. *Quaternary Res., 44*, 438–446.
5. Chaloner, W. G., & Creber, G. T., (1990). Do fossil plants give a climatic signal? *J. Geol. Soc. London, 147*, 343–350.
6. Chaloner, W. G., & McElwain, J., (1997). The fossil plant record and global climatic change. *Rev. Paleobot. Palynol., 95*, 73–82.
7. Davies-Vollum, K. S., (1997). Early Paleocene paleoclimatic inferences from fossil floras of the western interior, USA. *Paleogeogr. Paleoclimatol. Paleoecol., 136*, 145–164.
8. Dilcher, D. L., (1974). Approaches to the identification of fossil leaf remains. *Bot. Rev., 40*, 1–157.
9. Dilcher, D. L., (2000). Toward a new synthesis: Major evolutionary trends in the angiosperm fossil record. *PNAS, 97*, 7030–7036.
10. Dilcher, D. S., (1973). A Paleoclimatic interpretation of the Eocene Floras of Southeastern North America. In: Graham, A., (eds.), *Vegetation and Vegetational History of Northern Latin America* (pp. 39–59). Amsterdam: Elsevier. Chapter 2.
11. Edwards, T. W.D., Graf, W., Trimborn, P., Stichler, W., Lipp, J., & Payer, H. D., (2000). $d^{13}C$ response surface resolves humidity and temperature signals in trees. *Geochim. Cosmochim. Acta, 64,* 161–167.
12. Francis, J. E., & Poole, I., (2002). Cretaceous and early Tertiary climates of Antarctica: Evidence from fossil wood. *Paleogeogr. Paleoclimatol. Paleoecol., 182*, 47–64.
13. Givnish, T. J., (1987). Comparative studies of leaf form: Assessing the relative roles of selective pressures and phylogenetic constraints. *New Phytol., 106*, 131–160.

14. Greenwood, D. R., & Wing, S., (1995). Eocene continental climates and latitudinal temperature gradients. *Geology, 23*, 1044–1048.
15. Greenwood, D. R., (1992). Taphonomic constraints on foliar physiognomic interpretations of Late Cretaceous and Tertiary paleoclimates. *Rev. Paleobot. Palynol., 71*, 149–190.
16. Gregory-Wodzicki, K. M., (2000). Relationships between leaf morphology and climate, Bolivia: Implications for estimating paleoclimate from fossil floras. *Paleobiology, 26*, 668–688.
17. Halloy, S. R.P., & Mark, A. F., (1996). Comparative leaf morphology spectra of plant communities in New Zealand, the Andes and the European Alps. *J. R. Soc. N. Z., 26*, 41–78.
18. Jacobs, B. F., (2002). Estimation of low-latitude paleoclimates using fossil angiosperm leaves: Examples from the Miocene Tugen Hills, Kenya. *Paleobiology, 28*, 399–421.
19. Kowalski, E. A., & Dilcher, D. L., (2003). Warmer paleotemperatures for terrestrial ecosystems. *PNAS, 100*, 167–170.
20. Kowalski, E. A., (2002). Mean annual temperature estimation based on leaf morphology: A test from tropical South America. *Paleogeogr. Paleoclimatol. Paleoecol., 188*, 141–165.
21. Kurschner, W. M., Stulen, I., Wagner, F., & Kuiper, P. J.C., (1998). Comparison of paleobotanical observations with experimental data on the leaf anatomy of Durmast Oak [*Quercus petraea* (Fagaceae)] in response to environmental change. *Ann. Bot., 81*, 657–664.
22. Martin, H. A., (1997). The use of ecological tolerances for the reconstruction of Tertiary paleoclimates. *Aust. J. Bot., 45,* 475–492.
23. Mosbrugger, V., & Utescher, T., (1997). The coexistence approach: A method for quantitative reconstructions of Tertiary terrestrial paleoclimate data using plant fossils. *Paleogeogr. Paleoclimatol. Paleoecol., 134*, 61–86.
24. Parkhurst, D. F., & Loucks, O. L., (1972). Optimal leaf size in relation to environment. *J. Ecol., 60*, 505–537.
25. Roth, J. L., & Dilcher, D. L., (1978). Some considerations in leaf size and leaf margin analysis of fossil leaves. *Cour. Forsch. – Inst. Senckenberg, 30*, 165–171.
26. Royer, D. L., (2001). Stomatal density and stomatal index as indicators of atmospheric CO_2 concentration. *Rev. Paleobot. Palynol., 114*, 1–28.

27. Sinka, K. J., & Atkinson, T. C., (1999). A mutual climatic range method for reconstructing paleoclimate from plant remains. *J. Geol. Soc. London, 156*, 381–396.

28. Spicer, R. A., (1989). The formation and interpretation of plant fossil assemblages. *Adv. Bot. Res., 16*, 95–191.

29. Stranks, L., & England, P., (1997). The use of a resemblance function in the measurement of climatic parameters from the physiognomy of woody dicotyledons. *Paleogeogr. Paleoclimatol. Paleoecol., 131*, 15–28.

30. Sun, B., Dilcher, D. L., Beerling, D. J., Zhang, C., Yan, D., & Kowalski, E., (2003). Variation in *Ginkgo biloba* L. leaf characters across a climatic gradient in China. *PNAS, 100*, 7141–7146.

31. Terral, J.-F., & Mengal, X., (1999). Reconstruction of Holocene climate in southern France and eastern Spain using quantitative anatomy of olive wood and archeological charcoal. *Paleogeogr. Paleoclimatol. Paleoecol., 153*, 71–92.

32. Uhl, D., & Mosbrugger, V., (1999). Leaf venation density as a climate and environmental proxy: A critical review and new data. *Paleogeogr. Paleoclimatol. Paleoecol., 149*, 15–26.

33. Uhl, D., Walther, H., & Mosbrugger, V., (2002). Leaf venation density in angiosperm leaves as related to climatic and environmental conditions problems and potential for paleoclimatology. *N. Jb. Geol. Paläont. Abh., 224*, 49–95.

34. Wiemann, M. C., Dilcher, D. L., & Manchester, S. R., (2001). Estimation of mean annual temperature from leaf and wood physiognomy. *Forest Sci., 47*, 141–149.

35. Wiemann, M. C., Manchester, S. R., & Wheeler, E., (1999). Paleotemperature estimation from dicotyledonous wood anatomical characters. *Palaios, 14*, 459–474.

36. Wiemann, M. C., Wheeler, E., Manchester, S. R., & Portier, K. M., (1998). Dicotyledonous wood anatomical characters as predictors of climate. *Paleogeogr. Paleoclimatol. Paleoecol., 139*, 83–100.

37. Wilf, P., (1997). When are leaves good thermometers? A new case for leaf margin analysis. *Paleobiology, 23*, 373–390.

38. Wilf, P., Wing, S., Greenwood, D. R., & Greenwood, C. L., (1998). Using fossil leaves as paleoprecipitation indicators: An Eocene example. *Geology, 26*, 203–206.

39. Wing, S., & Greenwood, D. R., (1993). Fossils and fossil climate: The case for equable continental interiors in the Eocene. *Phil. Trans. R. Soc. Lond. B, 341*, 243–252.
40. Wolfe, J. A., (1993). A method of obtaining climatic parameters from leaf assemblages. *USGS Bull., 2040*, 71.
41. Woodcock, D. W., & Ignas, C. M., (1994). Prevalence of wood characters in eastern North America: What characters are most promising for interpreting climates from fossil wood? *Am. J. Bot., 81,* 1243–1251.

Chapter 4

Paleocean Modeling and Paleoceanography

CONTENTS

4.1. Introduction 112
4.2. Paleocean Modeling Challenges 113
4.3. Setting Up Deep Paleocean Model Investigations 113
4.4. Prognosis of Paleocean Modeling 116
4.5. Meridional Overturning Circulation (MOC) and Deep-Water Formation 118
4.6. Heat Transport and Ocean Gateways 119
4.7. Paleoceanography 121
4.8. Paleoceanographic Techniques 122
4.9. Paleoceanophic Processes 125
4.10. Paleocene-Eocene Thermal Maximum (PETM) 128
References 137

4.1. INTRODUCTION

The application of numerical or analytic techniques for the estimation and understanding of the climate, biogeochemical interactions, and the ocean's past circulation is known as paleoceanographic (paleocean) modeling. Paleocean modeling is generally a mechanism for formulating concrete estimations testable opposes to paleoenvironmental proxies, and for finding the mechanistic operation of the ocean (Poulsen et al., 1998, 2001, 2002).

For studying the particular questions, paleocean modeling is used, such as: (i) To which extent, the leading oceanographic attributes are sensitive, for example, the meridional overturning circulation (MOC) or the Gulf Stream to the changes in boundary condition? (ii) What is the tracer distribution and current structure of a given prior time interval? (iii) Define the participation of the ocean to control the response of the Earth systems to greenhouse gas (GHG) forcing, and also explain the role of the ocean in GHG repository and discharging by the history of the Earth? (iv) In what way the modes of variability, e.g., interdecadal, millennial-scale variability, and interannual, may change on various conditions? By this some more common questions might be answered as well, like: (v) To understand the change in climate, to what extent the changes in ocean heat transport (OHT) are significant, and how sensitive is the OHT: multiple equilibria, gateways, GHGs? Because of the recent improvements in modeling, currently, answers to such questions are pending till the development of a suitable framework and data that can be able to interpret them (Figure 4.1).

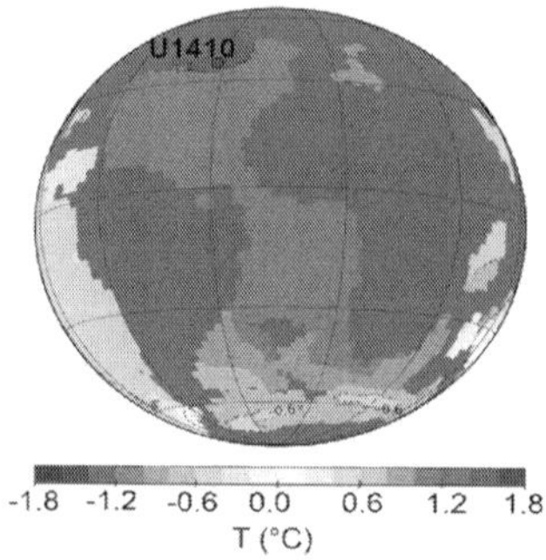

Figure 4.1. Boreal winter paleoceanography that is simulated using the COSMOS model.

Source: https://www.researchgate.net/figure/Boreal-winter-paleoceanography-as-simulated-by-the-COSMOS-model-Seasonally-averaged-Sea_fig4_322244387.

4.2. PALEOCEAN MODELING CHALLENGES

The ocean is connected to climate by its surface: through heat, fluxes of sunlight, momentum, and water, and through its state, that is sea surface temperatures (SSTs). Moreover, it is biogeochemically connected to the biosphere, via carbon dioxide, dust, and fluxes of sunlight, and indications of inorganic and organic chemicals.

Hence, it is needed to force the fluxes via-and state of-the interface within the atmosphere and ocean in order to estimate its role in biogeochemical cycles and climate and to perceive the ocean's phenomenology and dynamics (Power and Kleeman, 1993, 1994; Seager et al., 1995).

As the ocean is compelled, to first order, by its top boundary-the constraints of which are weakly recognized in the past-paleocean modeling demands either the quantitatively predicted such surface conditions, or for best case, prognosed objectively inside a coupled model of ocean-atmosphere. Moreover, there are just poor conditions on paleobathymetry, for paleo-oceanographic modeling a significant boundary constraint (Crowley, 1998), and horizontal and vertical ocean current velocities that may be used being a trial of model estimations are recognized poorly.

4.3. SETTING UP DEEP PALEOCEAN MODEL INVESTIGATIONS

Except for the usual challenges of ocean modeling (for instance, Gent et al., 1998), there are three challenges that are facing by the deep past paleoceanographic simulations: boundary conditions, model spin up, and uncertainty produced by the selection of initial conditions.

4.3.1. Initial Conditions

For experiments modeling, proxies that may be utilized as initial conditions are distributed sparsely in space and time, and they have deep-rooted uncertainties connected to them, and may not restrain significant quantities (for example, salinity). Usually, this challenge is overwhelmed by employing outcome from an atmospheric general circulation model (AGCM), or by determining an initial SST distribution introduced from proxies, which consequently has estimated SSTs, considering some distribution of heat transport and oceanic mixed layer depth. Generally, the surface salinities are dependent upon the result given by such AGCM simulations later considering an empirical association within surface salinity and precipitation subtracted evaporation (P-E) distributions, or rather, a global constant value might be given as an initial condition.

4.3.2. Boundary Conditions

The way of passing the moisture, momentum fluxes, and heat to the ocean model, and also the static features like bathymetry are included in these during the whole simulation. Conventionally, fluxes have been controlled by the usage of so-called Haney restoring of temperature (Haney, 1971) and the requirements of net virtual salt fluxes, which is considered to be a mixed boundary condition. Normally, the "weaker" restoring is closer to this boundary condition resembling atmospheric interaction, although mixed boundary conditions have been revealed to be exposed to spurious multiple equilibria (Weaver and Sarachik, 1991; Saravanan and McWilliams, 1995; Power and Kleeman, 1993, 1994). When the coupled model is not utilized, then the usage of so-called bulk forcing (Doney et al., 1998; Brady et al., 1998), or rather the addition of even a simple depiction of atmospheric fluxes as in Najjar et al. (2002) and Nong et al. (2000), is the best. Generally, the surface wind stress patterns are particularized dependent upon a theoretical distribution or contemporary conditions, or formed by an AGCM.

As a non-interactive atmosphere is intrinsically unrealistic, atmosphere like where temperature, freshwater fluxes, and winds cannot tailor to altering ocean conditions, in most situations, rather than an uncoupled model, a coupled model is favored, even though the focus is rigidly oceanographic (Doney et al., 1998). The more realistic pattern of flux variations and surface state is also allowed, and concerning OHT to be a well-posed prognostic variable, which is not permitted when surface temperature patterns are determined (Nong et al., 2000; Seager et al., 1995; Huber et al., 2003a, b).

A boundary condition that has often been a topic of interest in the design of paleocean modeling experiments is known as ocean bathymetry (Bice et al., 1997, 1998). Eschewing bathymetric variation and a flat bottom configuration altogether have been used by some researchers (Bush and Philander, 1997; Barron and Peterson, 1991) and it is suggested by the dynamical arguments that most ocean transports and current patterns are not receptive to bathymetric variations (summed up in Bush, 1997). However, the analysis of some currents absolutely includes bathymetry (Bice et al., 1997, 1998). Possibly more chiefly, the extent of upwelling by the abyss is controlled to a large extent by the existence of abyssal bathymetric increase (e.g., the flow of the geostrophically balanced meridional along ocean ridges) and hence, some scheme is possibly valuable even concerning the gross features regarding the ocean circulation (for example, Toggweiler and Samuels, 1995).

4.3.3. Deep Ocean Spin Up Process

Integration of the complete system to a quasi-steady state is another challenge, inclusion of deep ocean while doing paleocean modeling studies. However, ocean models are computationally low-cost compared to their atmospheric cousins, at least the models of non-eddy resolving used in paleoclimate and climate studies. Now it is possible, in the studies of ocean-only ocean general circulation model (OGCM), to combine the model to equilibrium exclusive of so-called deep ocean acceleration, even though, while spin up, accelerating the deep ocean tracer fields has been demonstrated to expedite equilibration exclusive of unduly influencing the final state of climate (Danabasoglu, 2004; Danabasoglu et al., 1996). Nevertheless, as mentioned above, for the purposes of climate, a coupled model is most recommended, therefore, salinity fields, surface temperature, freshwater transport, and ocean heat are completely prognostic. A "drift" of climate is displayed historically by several coupled models away from contemporary values, and even on contemporary conditions, which have inhibited them for usage in paleoclimatology, although this challenge of modeling has been reduced in the current models' generation. For present computational resources, it has not been possible to pursue the perfect solution using a coupled general circulation model (CGCM), which is responsible to combine many multi-millennial simulations. There are two options: (i) use a method to accelerate the convergence of an entire CGCM, or (ii) use a simplified coupled model having less expensive computationally. Currently, a lot of "intermediate" complexity models or simplified CGCMs exist, and generally, in the modeling toolbox, they are considered to be a valuable tool (Saravanan and McWilliams, 1995; Claussen et al., 2002; Weaver et al., 2001; Poussart et al., 1999). For insufficiency of such tools, repeated accelerated spin up methods have been introduced to permit deep ocean simulation of many 1,000 years to be accomplished in just surface ocean computation of many 100 years with entirely coupled GCMs.

There are several techniques-many of them exploit two extant processes: deep ocean acceleration, and iteration of uncoupled and coupled modeling phases. Several researchers have applied and validated iteration technique in considerably distant forms (Kutzbach and Liu, 1997; Sausen and Voss 1996; Vavrus et al., 2000; Liu et al., 1999, 2004). A different technique is described by one of the scientists that use deep ocean acceleration openly in the completely coupled integration and in the Eocene, this has been demonstrated to offer convergent climates exploiting the iterative technique, hence, any of the family of methods is considered to be a potential candidate for usage (Shin et al., 2003). No guarantees are given for the generalization of it for various model configurations or models, or that some acceleration technique will employ evenly well at any time interval. Thus, a primary need of any approach is the basic convergence validation of some modeling framework over a reference condition (supposedly the extant) to some suitable similarity level (Huber and Sloan, 2001).

4.3.4. Modeling Choices

There is no ideal model, and several models and parameterizations in extant are improved for a few intentions than others. However, there are some generic directions, somewhat with complication, GCM caliber models, that must be assumed. One of the most contentious challenges in modern oceanography and most significant is the oceanic vertical diffusion. Probably, it is to have been indeed more significant in the former times (Otto-Bliesner et al., 2002; Liu et al., 2004; Huber and Sloan, 2001). The most debated topics are the physical procedures that give rise to it beyond this background value, and the certain "background" diffusivity value. For an approximate of such background value, for the existence of some spread, 0.1 cm^2 s^1 is a credible noticed number, and in applications of paleoceangraphic, there must be the default values in this range. The generally used values are 10–100 times higher than these are used indeed in modern simulations, which focuses on the lacks of those models. The addition of advanced treatments of isopycnal eddy and diapycnal mixing is significant, and extant it seems that an alliance of the Gent-McWilliams and KPP treatments of such procedures is best from the perspective of the tracer characteristics and tracer circulation (Gent et al., 1998). For such parameterizations, there is significant space for the betterment and ideally, many vertical diffusion techniques must be tested as there may be fundamentally different resulting ocean circulations (for example, Nilsson et al., 2003).

4.4. PROGNOSIS OF PALEOCEAN MODELING

Horizontal ocean currents. OGCMs and also a host of simplified models predict them. They are the values that are much firmed with the well-represented and well-understood ocean dynamics. Generally, it is not important for an OGCM to estimate velocities. The compelling wind fields are reflected largely by the upper ocean flows and surface, and in turn, leading to "shallow water" ocean models, or indeed analytic computations like the "island rule" or the predictions of the disposition, shape, and strength of gyre circulations and also the currents of the western boundary in equilibrium with them are allowed by the Sverdrup transport for some wind field (Nof, 2000; Huber and Nof, 2006; Nof and Van Gorder, 2002). Within the deep ocean, features are reproduced mainly by the outcomes of the OGCM, which are adducible by Stommel-Arons theory. Using the records of the paleocirculation direction and intensity, these estimates may be validated (Figure 4.2).

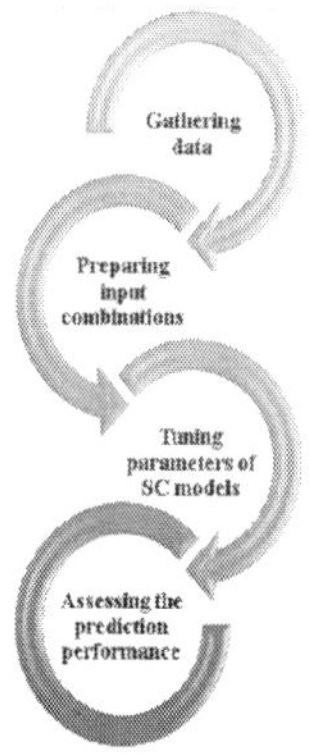

Figure 4.2. Visionary plan of soft computing models for the prediction of sea surface temperature.

Source: https://progearthplanetsci.springeropen.com/articles/10.1186/s40645-020-00400-9.

Unluckily, there are weakly constrained current directions, and in "deep-time" paleoceanography, current magnitudes are amidst the much poorly recognized features, because direct proxies do not strongly constrain such features, and there are qualitative indirect proxies (McGowran et al., 1997; Carter et al., 1996). However, valuable work has been done in estimating directions of paleocean current and verifying these upon proxy understanding. Some remarkable studies involve those in the global ocean, in the Cretaceous Tethyan current regime (Bush and Philander, 1997; Barron and Peterson, 1991), the Permian (Winguth et al., 2002), and in the Turonian seaway on a regional scale (Slingerland et al., 1996). An effort has been made, in more late periods, to figure out alterations in the direction of flow within the Pacific and Atlantic, i.e., over the Miocene (Nisancioglu et al., 2003; Nof and Van Gorder, 2002; Omta and Dijkstra, 2003), even though in such studies, there is a lack of the comparison of explicit model-data. In such studies, for the comparison of proxy records, there would be considerable assistance of the dispersal models and the more widespread usage of biological and sediment transport. A scientist has taken the initial steps towards this direction Huber et al. (2004).

A feature, recognized as upwelling, is more tightly fixed to the proxy record as compared to the direction of the ocean current as it is connected more directly to a noticeable quantity (burial and productivity). Hints about the structure of circulation changes, changes in the global cycling of chemical species whose encounter is possible, and primary shifts of productivity (Moore et al., 2004) can be provided by the understanding the way by which upwelling patterns may have altered in the past (Toggweiler and Bjornsson, 2000). Over these time intervals, for putting constraints upon organic carbon burial rates, particularly, upwelling predictions may be valuable (Handoh et al., 1999, 2003).

4.5. MERIDIONAL OVERTURNING CIRCULATION (MOC) AND DEEP-WATER FORMATION

A significant and complicated process is the change of surface waters into deep water, and intermediate and shallow mode, and successive recycling of the particular waters into the surface. In the ocean, it is vital to the redistribution of nutrients, heat, and other tracers and is, therefore, of high interest in paleoceanography (Figure 4.3) (Bice and Marotzke, 2002).

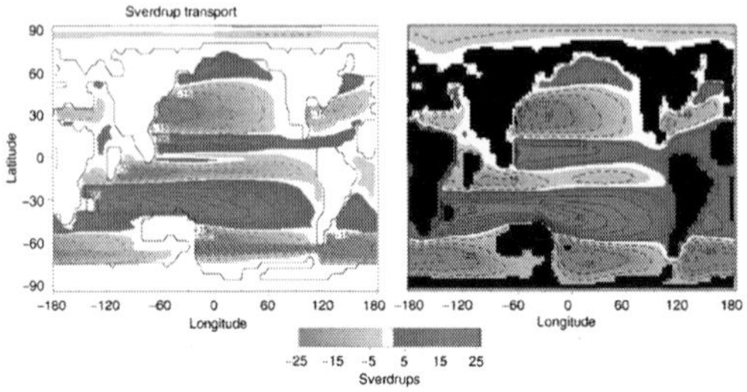

Figure 4.3. Barotropic stream function generated by the ocean model (see right figure), and the general and theoretical wind-driven circulation model using ocean circulations (refers to left figure).

Source: https://link.springer.com/referenceworkentry/10.1007%2F978-1-4020-4411-3_172.

Note: (Right) Counterclockwise gyres are shown in red, and clockwise gyres are shown in blue. At the interval of 10 Sverdrup, the flow is contoured. The equivalence within the upper ocean velocities indicates openly that-for the well-known wind fields, or clearly recommended-this distribution will completely control the primary ocean current systems. (Left) The predicted wind-driven ocean gyre transport having the AGCM-derived winds evaluated from the Sverdrup transport, in equilibrium, forced using the wind stress curl. The first order is suggested by the theory that barotropic velocities must conform approximately to the Sverdrup transport.

Similar to the fragile data-based constraints on the velocity of the ocean, there are fragile constraints on its integral characteristics, like the meridional overturning stream function. The computation of spatial and vertical gradients of $d^{13}C$ is one of the generally used constraints that is an isotopic tracer corresponding to the carbon cycle and nutrient distribution (for example, Corfield and Norris, 1996). Where there is an existence of enough $d^{13}C$ gradients, there can be a determination of the position of areas of deep-water formation and there might be the calculation of bulk flow patterns. The latest novelty that may include considerable detail to the depiction that has been introduced from $d^{13}C$ tracer distributions is the usage of other water mass tracers, like neodymium (for example, Thomas, 2004; Hotinski et al., 2001).

4.6. HEAT TRANSPORT AND OCEAN GATEWAYS

Previous changes of climate have been often associated with the changes of ocean gateway, e.g., closure of the Panamanian Gateway and the commencement of the Drake Passage (Kennett, 1977; Lawver and Gahagan, 1998). This thought meets its support in the estimated correlation between climatic change and major gateway changes. The heat transport, overturning circulation of the ocean, and consequent temperature distribution have been demonstrated by several modeling studies to be sensitive to some extent to ocean gateways changes (Mikolajewicz et al., 1993, 1997). Moreover, there is a solid line of modeling and theoretical proof that supports the theory which is stated that the changes in the past in this upwelling might have been significantly important at both low latitudes (Hotinski and Toggweiler, 2003) and high latitudes and that in the Southern Ocean, the deep-water wind-driven upwelling has a vital role in contemporary ocean circulation (Toggweiler and Bjornsson, 2000). The most often cited instance of a direct influence of changes of the gateway on OHT is because of the formation of the Antarctic circumpolar current (ACC). It is observed in a study that the formation of the ACC results in a warming up of the Northern Hemisphere, and a tempering down of the Southern Hemisphere as the flows connected to the ACC "steal" warm over the equator.

As a similar prognosis is presented by all of the above-mentioned studies according to the responses of the physical oceanographic to changes in the gateway, the response of surface temperature, such as the climatic effect of this change demonstrates a vast spread. In researches where a simple depiction of coupling of atmosphere-ocean exchanges are included, closing, and opening of Southern Ocean gateways turn into the changes in Southern Ocean surface temperature from a low value, i.e., 0.80°C (Mikolajewicz et al., 1993), to a midst range, i.e., 1.50°C, and a high value, i.e., 3.50°C. While using a CGCM, the consequent change in the Southern Ocean is in the midst of the range of former calculations. The differences in the regimen of the "top" boundary condition are represented by the spread of values, counting damping through the atmosphere and feedbacks of sea ice, instead of exhibiting an effective difference in estimated behavior of ocean dynamics (Figure 4.4) (Nong et al., 2000; Huber et al., 2003b).

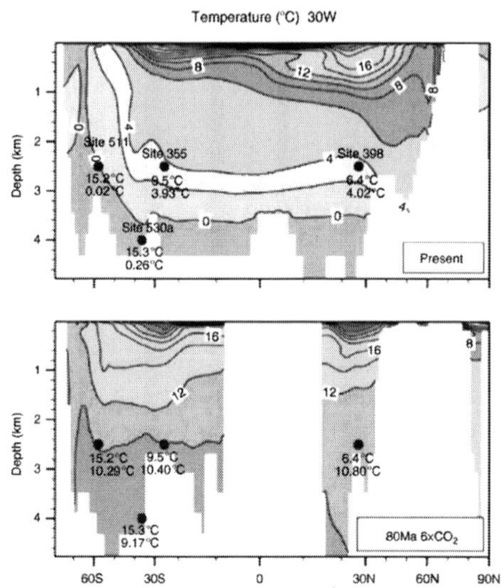

Figure 4.4. Cretaceous paleotemperature predictions presented from DSDP and ODP cores dependent upon deep-dwelling fauna are inadequate and show a broad range of values, but fix to a too warmer deep ocean.

Source: https://agupubs.onlinelibrary.wiley.com/doi/full/10.1029/2001JD0008 21.

Note: Ocean temperatures are predicted by the NCAR Climate System Model for a cut across in the Atlantic Ocean lower than 2.50 km depth of around 10.0°C (i.e., bottom values) comparable to proxy approximates of 6.0–15.0°C (i.e., top values). It is indicated by the model simulation that the origin of the warmer deep ocean temperatures is settling in the warmer southern polar waters (Otto-Bliesner et al., 2002)

4.7. PALEOCEANOGRAPHY

The study of the past of the oceans is known as paleoceanography. Facets of climatology, oceanography, chemistry, geology, and biology are incorporated. Inorganic marine sediments, and biogenic, and also corals, are the principal sources of knowledge. Benthic fossils and planktonic are included in biogenic sediment, on the other hand, dust, and ice-rafted debris are included in inorganic sediment. The main sources of proxy data on land are the erosional and paleo-shorelines features, and also paleomarine sediments' outcroppings. Information regarding paleoceans can be given indirectly by the glaciological records as well. A significant role is given by the capacity of the ocean's high heat and its competency to transit energy and to release and sequester GHGs, in support to find out the planet's climate state. Hence, paleoceanographic research and paleoclimatology are linked closely as well (Figure 4.5).

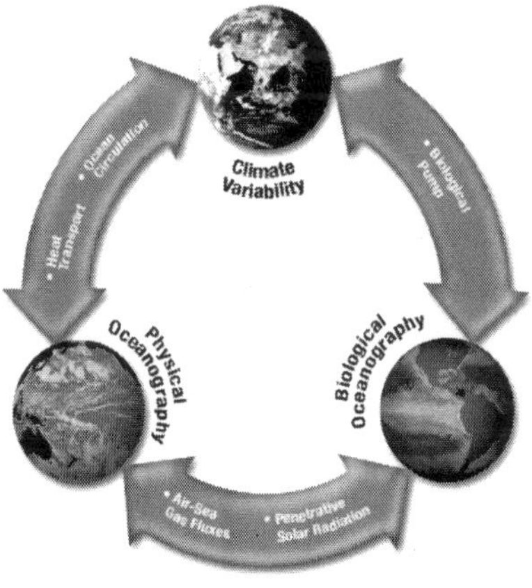

Figure 4.5. Climate change and oceanography.

Source: https://science.nasa.gov/earth-science/oceanography/ocean-earth-system/climate-variability.

4.8. PALEOCEANOGRAPHIC TECHNIQUES

Climatic detective study is required by the reconstruction of paleocean dynamics and properties. The understanding and dating of paleoclimatic records are involved and also the description of dynamical and physical constraints which determine viable characteristics and circulation patterns.

4.8.1. Reconstructions (Proxy Data)

Direct evaluation of the amount of interest towards oceanographers only expands into the comparably late past and in many situations does not go farther back than the halfway to 19th century. There is a reliance on indirect evidence to work on the ocean in the time of the periods that have no direct calculations. Being a data source, historical documents can be utilized. Strengths of previous prevailing winds and directions estimates have been provided by the sailing times and ship logbooks over often travelled passage (Brazdil et al., 2005). Since the 1500s, the intensity and frequency of El Niño events have been constructed again dependent partially upon historical accounts of crop losses and abundant floods (Quinn and Neal, 1987). Past qualitative explanations are provided by this kind of analysis.

The proxy makes possible the quantitative reconstructions in which a quantity can be calculated and is kept in a natural collection, viewed as a stand-in of the parameter of interest. It is a fundamental need that the connection of the quantity of interest and the proxy parameter must be known. For such cases, the past of the proxy variable can be transformed into the past of the interested variable using mathematical expressions like:

$$\mathrm{Int}_t = f(Prx_t) \qquad (1)$$

which shows that Int, which is the parameter of interest is a function of Prx, i.e., proxy quantity. The index t indicates the time. Such a type of equations is generally known as transfer function. The confidence using which Int can be measured will be based on a sequence of factors, initiating from the quality of the proxy calculation. Also, a typical source of uncertainty, and pertinent, is how well f refers to the relationship within Int and Prx (Werner et al., 2000; Schmidt, 1999; Butzin et al., 2005; Kaplan et al., 2002).

In many cases, with the comparison of a relevant proxy data set and the directly calculated values of the quantity of interest, transfer functions are empirically found. It is possible that the function attained using this method is not general which is a probable source of error, but actually, it can indicate a relationship within Int and Prx which is unique to the datasets utilized to produce f. With the expansion of the temporal and spatial coverage of the data which is exploited to make the transfer function, this problem can be overcome. Still, even supposing f to be an ideal indication to represent the present relationship of the quantity of interest and the proxy with each other, there is no assurance of the past connection between them (Paul and Schäfer-Neth, 2005). Following are the descriptions of various types of proxies.

The methodical usage of proxies within quantitative reconstructions of previous oceanic surroundings has started since the second half of the 20th century. After that, a large number of proxy methods have been introduced and many of them are still in the process of development. There are six generous classes of the proxies based on the kind of direct measurement (Fischer and Wefer, 1999). These proxies along with the concise explanations of their chief variables of interest are given below. The proxies related to the fields, e.g., paleoproductivity, paleocirculation, and paleotemperature, are presented in detail in the following chapters. The literature presents a thorough analysis of oceanographic proxies (Bradley, 1999; Henderson, 2002).

4.8.1.1. Microfossil Assemblages

The comparative abundance of benthic and planktonic species of coccoliths, foraminifera, diatoms, radiolaria, and other organisms can be utilized to measure previous ocean temperature, sea ice distribution, and productivity. This type of proxy is utilized for the CLIMAP project, which generated the initial global distribution of SST for the latest glacial maximum (CLIMAP-Project Members, 1976).

4.8.1.2. Stable Isotopes

Stable isotopes are dependent upon the proportion within various isotopes of an element. Usually, the ratios are regulated with a reference value and recognized as a title suggested by the heavier isotope. For instance, the ratio within ^{16}O and ^{18}O is depicted by $d^{18}O$. Chiefly, the isotope readings are accessed from organic matter, foraminifera skeletons (i.e., tests), or alternate sources (for example, molecules of water in continental ice sheets). Organisms, e.g., corals, and foraminifera incorporates with the quantity of ^{18}O and grows with the decline of temperature. As compared to seawater, continental ice is correspondingly reduced in ^{18}O. This forms $d^{18}O$ which is a proxy for both the range of continental ice sheets and temperature. For pH,

d^{11} B is utilized being a proxy. From $d^{13}C$ and d^{15} N, nutrient concentration, past circulation, and productivity can be reconstructed (where ^{12}C is engaged with slender priority to ^{13}C while photosynthesis). Along with microfossil assemblages, the $d^{13}C$ and $d^{18}O$ are considered to be the paleoceanographic proxies along with the very widespread usage.

4.8.1.3. Radiogenic Isotopes

The flux of particles and the deep-water flow rate estimations from the water column to the sediments can be made by two products of their naturally hitting decay, protactinium (Pa) and thorium (Th), and the various solubilities of uranium. Being an estimate of productivity, this flux can be utilized as well. For the estimations of the age difference within deep waters and near the surface, ^{14}C is used which is kept in organic matter and thus, ventilation rates occur.

4.8.1.4. Biogenic Compounds

For the past productivity estimations, the concentrations of some compounds, particularly calcium carbonate, opal, and organic carbon are utilized. Calcium carbonate is considered to be the calcite compensation depth indicator as well. Protracted chained organic molecules, alkenones, are immune to degradation. There are two or three double bonds in the architecture of the alkenones formed from some coccolithophors. The temperature is reflected at the time of synthesis by the ratio within molecules having two and three double bonds.

4.8.1.5. Elements

The ratios and concentrations of specific elements in the organic remnant, sediment, corals, and tests are utilized as proxies as well. In biologically triggered marine carbonates, the ratios of magnesium to calcium (i.e., Mg/Ca) and strontium to calcium (i.e., Sr/Ca) are dependent on temperature. For nutrient reconstructions, there is a usage of a ratio of cadmium to calcium (Cd/Ca). The Ba/Ca ratio is the proxy of alkalinity, while barium concentration is the proxy for productivity.

4.8.1.6. Sedimentology

Qualitative information regarding bottom current speeds can be provided by the grain size distribution which acts as an ice-rafted debris indicator.

Rhythmites refer to the data regarding past tides that can be derived from layered sediments. For the establishment of the source areas and the transport direction as well for both wind and water-borne sediments, the mineralogy of the sediments can be utilized (Fischer and Wefer, 1999).

4.8.2. Reconstructions (Models)

Climate models and quantitative ocean are used by the theoretical way to paleoceanography to analyze observations and reconstruct ocean's paleoconditions. In the past, coupled climate models and various oceans have been utilized. Hither, a comprehensive overview of the hierarchy is given along with the usage of such models in the vast research area referring to paleoceanography (Clark et al., 2002, 2005; Bond et al., 2001).

Obviously, it is not possible to make an accurate model regarding the climate system; the momentum and position of around $5–10^{46}$ molecules, for only the ocean, must have to be measured at each moment of time. Otherwise, the atmosphere, ocean, land surface, or sea ice is divided into individual macroscopic elements having measurable properties like velocity, density, temperature, etc. The physical laws are followed by the condition and swap between such discrete elements and hence, numerical models can be used to find them. Also, the large-scale pattern can be affected by small scale procedures in each element, and thus, they must be parametrized (Meissner et al., 2003; Stocker and Wright, 1996; LeGrande et al., 2006; Crucifix, 2005; Roche et al., 2004).

4.9. PALEOCEANOPHIC PROCESSES

This section presents a brief overview of the methods and boundary conditions that might have, in the past, affected the oceans.

4.9.1. Paleotides

At the geological time, the history of tides is linked with the structure of ocean basins and the development of the Earth-Moon system. Friction is produced by the tidal currents at the base of the ocean leading to the angular momentum and the energy transfer connected to the Earth's rotation towards the Moon's orbital motion. Consequently, the Earth's spin is gradually slow down and the radius of the Moon's orbit is increased. Some of the estimates state that at 620 million years ago (Ma), there have been around 22 h long days and the distance between Earth-Moon has been approximately 96% of its current-day value (Williams, 2000). For tidal (and Earth-Moon system), the chief proxy transforms at millions of years periods is dependent upon the tidal rhythmites study, laminated sediments whose removal is connected to the tidal currents (Figure 4.6) (Williams, 2000).

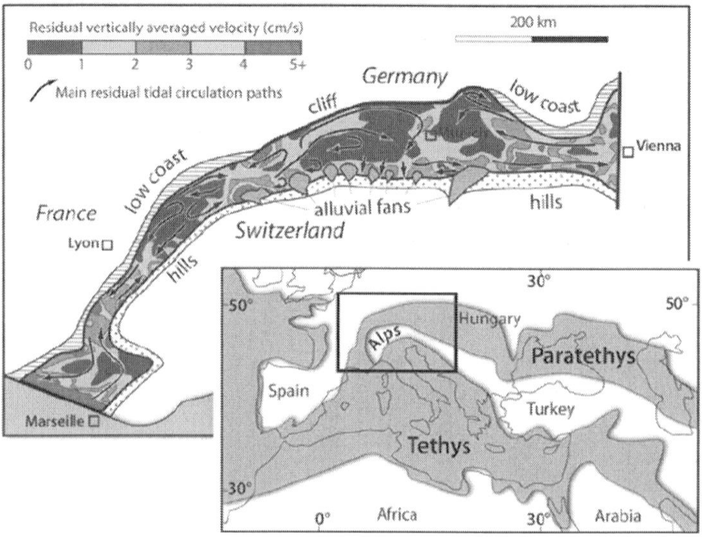

Figure 4.6. Paleotidal scheme of the Burdigalian Peri-Alpine Seaway.

Source: https://www.researchgate.net/figure/Paleotidal-modeling-of-the-Burdigalian-Peri-Alpine-Seaway-The-colors-show-the-velocity_fig18_234167661.

There is a strong dependence of tidal dissipation upon the structure of ocean basins. The influences of the tidal friction must be changed as the tectonic procedures make the change happens to the basins. A lot of clues indicate this case. For instance, the current dissipation rates are seen to be greater as compared to the mean rates at the history of the planet (Gills and Ray, 1999; Kagan, 1997).

The structure of the ocean basins, over the Quaternary, has been changed because of the absence or presence of ice shelves, and the sea-level changes. It is shown by the numerical models that while the last glacial, tidal heights in the Labrador Sea have been around twice as large in comparison with the current day conditions. This has brought about the indication that such higher tides could be the reason for the destabilization of the floating ice shelves and leads to the Heinrich events (Arbic et al., 2004). Another association between the climate and tides is related to a millennial tidal cycle. Very high tides have been presented, that has appeared after every 1800 years, can be the reason for cooling and increased mixing of the surface of the sea. This cooling would be linked to a sudden change in climate that is noticed with analogous periodicity (Keeling and Whorf, 2000). This hypothesis is contested by some scientists that argue about the weakness of the tidal forcing at such frequencies (Munk et al., 2001).

4.9.2. Radiation

The eventual source of energy is the incoming short-wave radiation coming from the Sun for the hydrological cycle, ocean dynamics, and life in the oceans. In the insolation intensity, seasonal, and geographic variations lead to pressure and temperature gradients having a significant effect on the system of climate and ocean dynamics. However, mostly the ocean surface circulation is wind-driven, while the winds themselves occur as the result of the unbalanced distribution of energy on Earth. Whereas, density-driven currents have a strong dependence on salinity gradients (which occur from evaporation/precipitation patterns and hence, are linked closely to the hydrological cycle) and temperature (and therefore, on energy distribution indirectly).

4.9.3. Changes in the Ocean Basin

For the ocean circulation, the boundary conditions are arranged by the structure of the ocean basins. Various flow patterns are resulted in from various configurations of the continent. Interpretation of how the boundaries have affected circulation patterns in the past might give a vision of the dynamics and other significant methods of the current day oceans. For instance, in the Tethys Ocean (160–14 Ma), the presence of a not obstructed small latitude passage has been linked to the raised heat transport towards poleward (Hotinski and Toggweiler, 2003). Obviously, in the past, both simulations and reconstructions of ocean currents have prevailed to several uncertainties. For instance, still, even there is no unequivocally acceptance of the presence of an outstanding, immense scale feature like the Tethys circumequatorial current (Barron and Peterson, 1989; Poulsen et al., 2001, 2002).

It is shown by many modelling and proxy studies that the closing and opening of paths within two basins can have an abundant influence on climate and ocean circulation. The closure of the Isthmus of Panama, considered to be the gap within the South and North American continents over 5–3 Ma, is assumed to have strengthened the Gulf Stream and the connected transport from northward of salt and heat towards the North Atlantic (Haug and Tiedemann, 1998). There is also an association of the closing of the Isthmus of Panama with the beginning of the Northern Hemisphere glaciation (Mudelsee and Raymo, 2005; Lear et al., 2003; Keigwin, 1982) and also with the North Pacific overturning circulation and the alterations in the properties of the water mass (Motoi et al., 2005). There is an association of the opening of the Drake pathway between South America and Antarctica (28–33 Ma) and the ensuing formation of the ACC with the Antarctic glaciation (Toggweiler and Bjornsson, 2000; Kennett, 1977). It has been presented that, during the Holocene, the corresponding balance of the Atlantic MOC is associated with the open connection within the Pacific and Arctic Oceans yielded by the Bering Strait (De Boer and Nof, 2004). In this case, the MOC is made unstable by the fewer salinity anomalies occurring in the North Atlantic for the blocked strait (as all along the last glacial). Whereas, the less salinity water flux by a broad open Bering Strait from the Pacific towards the Arctic all along the conditions having greater sea levels (latest interglacial, 115–130 ka BP) has been connected to the too unstable MOC (Montenegro et al., 2006; Shaffer and Bendsten, 1994).

4.9.4. Ice

For the past 55 million years, progressive cooling has led to the existence of vast continental ice sheets in both Hemisphere, i.e., Northern, and Southern. In North America and Eurasia, the intensification of glaciation of the Northern Hemisphere is taken place within 2.7–2.5 Ma. With rising terrestrial ice sheets, an intensification of climate oscillations is shown by sediment records between interglacial warm periods upon timescales and utmost cold glacial maxima corresponding to the orbital parameters. For paleoceans, the boundary conditions dramatically fluctuate within interglacial and glacial periods. In Eurasia and North America, in the melting and creation of vast ice sheets, the essential differences in shifts in the range of sea ice and atmospheric CO_2 concentrations have changed the salinity distribution, radiation balance, oceans productivity, and dynamical forcing.

4.10. PALEOCENE-EOCENE THERMAL MAXIMUM (PETM)

Globally around 55.5 Ma ago in terrestrial and marine sections, there has been a greatly abundant carbon isotope ($d^{13}C$) excursion (CIE) which is exclusive in the Tertiary based on its onset rapidity and size. Some scientists have initially acknowledged this event in multiple cores and are contemporaneous with an important warm anomaly (oxygen isotope anomaly as well) and hence, has been recognized as Paleocene-Eocene Thermal Maximum (PETM) (Kennett and Stott, 1991). Now, the foundation of this event is used to specify the Paleocene-Eocene transition boundary because of the faunal, ecological, and physical gradients crosswise, and the singular sort of this perturbation. The former description of the P/E boundary has been later undoubtedly (over the foundation of the Ypresian section at Europe), and therefore, established the CIE in the former Paleocene. Hence, earlier scientists have referred to this event being a Late (or recent) Paleocene Thermal Maximum (LPTM). Since some researchers have referred to it being an initial Eocene thermal maximum (IETM). Together with Clarkforkian/Wasatchian transition, the CIE is considered to be contemporaneous which is late recognized as the boundary of Paleocene-Eocene over the North American Land Mammal Series.

Using the frequent hiatuses obtained close to the Paleocene-Eocene transition and a number of not completed sections has handicapped the full characterization of the event. For this transition, the global stratigraphic section and point (GSSP) has been specified as a Dababiya Section, which is 35 km to the south of Luxor, over the right bank of the Nile Valley. Between reverse-polarity magnetic period Chron 24r and nano-Plankton zone 9 (NP-9), the event has occurred. It is estimated that the PETM dating is about 55.5 Ma, dependent on the chronology of employing interpolation within well-dated ash layers.

4.10.1. Background

The Paleocene has been a time of abundant shifts in the world's environment ensuing the interruptions that happened by the Cretaceous Tertiary boundary (65 Ma). This epoch is marked by rising ocean productivity and evolutionary rise in diversity, resulting in the remarkably positive values of the carbon isotope over the Tertiary.

By sub-tropical flora, and fauna (ferns, crocodiles, etc.), the polar regions (e.g., Ellesmere Island) are repleted, and generally, temperatures are warm. For vast glacial ice, there is no proof, however, at high northern latitudes, evidence for sea ice absence is not as persuasive. In contrast with the present, estimates of the tropical temperature do not demonstrate as much change (if any), but tropical proxies are recognized to suffer more from the probably complex factors corresponding to the tropical hydrological cycle and diagenetic changes. Concentrating more particularly on the 2 Ma recent Paleocene to previous Eocene transitional term, worldwide temperatures are raised (to an evaluated 2–4°C warmer as compared to the present day), deep ocean temperatures are raised by around $d^{13}C$ and 3°C values that are lower at this long period (Figure 4.7) (Aubry et al., 1996, 1998).

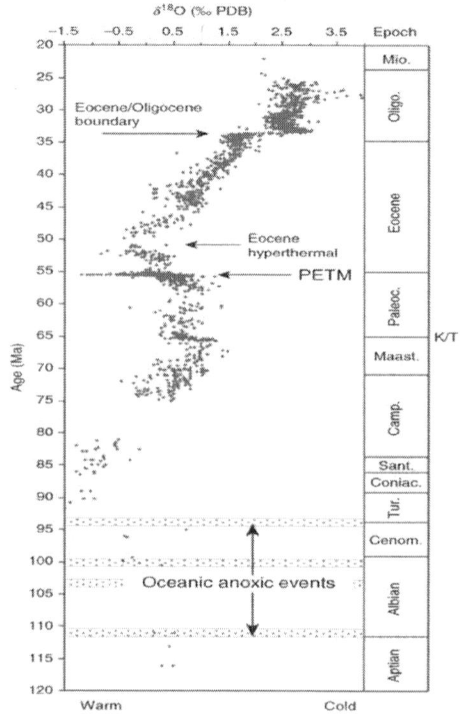

Figure 4.7. In marine carbonate sediments, the assemblage of $d^{18}O$ values as the cretaceous displaying the PETM irregularity in the context of Cenozoic variability (Zachos et al., 2001).

The Eocene Hyperthermal (or soon Eocene climatic optimum) is led which is the warmest term in the Cenozoic. During the whole term, the contact of the Asian and Indian plates has continued, however, there might already exist a land bridge in the late time of Maastrichtian. While the late Paleocene, at some point, there would have ceased island arc volcanism over this plate boundary. Although, the raised volcanism may accompany the opening of the North Atlantic, along with the extended ridge over there. Probably, a land bridge within Europe and North America is formed while this time period because of the quick extrusion of lava moves in the area of the Faroe-Rockall. There are abundant amounts of sediment in the Gulf of Alaska accretionary prism, that are being isolated (peaking within 65–55 Ma) (Hudson and Magoon, 2002). Climate might be affected by such tectonic procedures using the carbon cycle by worldwide temperatures using a restricting of volcanic aerosol input, taken out sinks of organic carbon, sea level and likely ocean circulation from the important gateways closing (especially the access to the Northern Atlantic and Tethys Ocean). Predominantly, over this period, the vegetation is sub-tropical and tropical along with climatic boundaries necessarily farther poleward instead of at present. It is estimated to have somewhat too extensive wetlands. For most of this period, similarities in intermediate and deep-water masses are directed to a sole, Southern Ocean source for deep-water (Corfield and Norris, 1998).

It is estimated to have somewhat high levels of CO_2 based on the models of geochemical cycling, however, from paleosol $d^{13}C$, there are several lines of conflicts of evidence, or stomatal density proxies based on the leaf, remnants a consensus elusive (Berner, 1994). Even the levels of alternate radiatively active elements (O_3, CH_4, volcanic aerosols) are very highly erratic. A long-term reduction of winds perhaps indicates the dustiness that shows up to have significantly reduced over this interval. Though, the connection within alterations in the long-term warming and radiative forcing, albeit discussed enough, remains unsubstantiated.

It should be clear that alterations at the complete boundary interval of Paleocene-Eocene (54.0–57.0 Ma) is significant and large, but not associated certainly with PETM event of too smaller duration which is discussed following.

4.10.2. PETM Event Characterization

The excursion of the carbon isotope is the unique and principle PETM definer. From when it has been found first, it is obtained in several terrestrial sequences and deep ocean cores. The excursion in the deep ocean is 3% to 4% and possesses a too quick onset (which is <10,000 yrs). It has been seen that the first excursions of carbon isotope have been very rapid and stepped, ensuing in bursts of below 1 kyr at this interval (Bains et al., 1999). In both paleosol carbonate sand mammalian tooth enamel (hydroxyapatite), it is measured that the terrestrial excursion is greater, i.e., up to 7% (Koch et al., 1992).

Though, the contemporaneous by this excursion which is somehow lightly lagged has an enormous rise in temperatures of deep ocean (up to 5°C to 7°C) as shown in carbonate sediments by $d^{18}O$ values. A similar alteration is also registered by the isotope signals originating from upper ocean-dwelling planktonic foraminifera, signifying the occurrence of warming at the full ocean column, especially at upper latitudes. However, tropical temperatures that are more unpredictable may have raised by 1°C to 2°C as well (Bralower et al., 1997). There has been a hypothesis of the changes of deep water that have been because of as witch by upper latitude deep-water generation to a tropical origin of warm saline bottom water (WSBW), however, still no unequivocal evidence is found for this (Kennett and Stott, 1991).

At this point, in carbonate dissolution, there also appears to have been an increase in the expansion (although not universal), obscure effort to encounter steady sequences over the event. At the CIE, in coastal areas, there is an inflation in the deposition of kaolinite clay which shows boosted weathering, or somewhat, increased transmission of weathered material towards the ocean. It has been seen by some coastal regions, inflation in the black shale's deposition, which represents increased productivity. At the Paleocene-Eocene transition, widespread discontinuities are appeared and also the proof of big slope failures over Blakes Nose is shown, in the North Atlantic especially (Figure 4.8) (Katz et al., 1999).

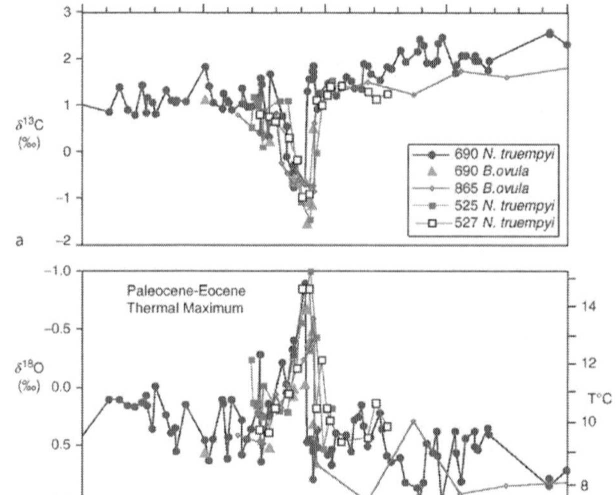

Figure 4.8. Thorough oxygen isotopic series and benthic foraminiferal carbon across the PETM occurrence from ODP stations 865 (Allison Guyot, Equatorial Pacific), 690 (Maud Rise, South Atlantic), 527 and 525 (Walvis Ridge, South Atlantic).

The benthic foraminiferal extinction (BFE) event coincidently appears at the onset of the excursion of carbon isotope, which has found the extant benthic species of around 50% vanish at that time. High changes in temperature, increased pH (constant with the raised carbonate dissolution), and a fall in dissolved O_2 has been connected to this extinction. Other benthic species groups (e.g., mollusks, ostracodes), inclusion with foraminifera, are undergone compelling turnover as well. There is a widespread increase of especially dinoflagellates and exotic planktonic taxa. The final appearance has been marked by the CIE on land, of many archaic mammal orders, and the commencement of a further "cosmopolitan" range of mammalian fauna in Eurasia and North America both. At this point, a massive ecological turnover is shown as well.

It is not easy to assess especially the period of any climate event vivid in the geological record. The remarkably valuable techniques depend upon the approximates of Milankovitch cycle variability (precessional cycles mostly 20 Kyr) in records of $d^{13}C$ which can be calculated at the event. Other values are based upon the rates of interstellar dust accumulation (as calculated by the ratios of $^3He/^4He$). The $d^{13}C$ excursion determines the best measurements for the period of the event that is within 90 to 200 Kyr. A compared to the estimates of the present day, these estimates are persistent for the perturbation's residence time to the carbon cycle. As the equilibrium of oceanic and atmospheric carbon occurs very much rapidly as compared to the carbon isotopes recycling time, climate perturbations will be smaller than the given timescale.

4.10.3. Methods of PETM Climate Change

For a primary cause, several candidates are eliminated by the support of the rapidity and size of the PETM CIE. A very massive input of depleted carbon is implied by a perturbation of up to 4% in the worldwide carbon cycle. From the valid measurements of the isotopic impression of terrestrial carbon, the required mass of carbon coming out of any source can be measured. For example, if the source is provided by the land biosphere, it would need to add three times the accumulation of carbon content of the present biosphere, and thus, at this time, no disruption on this type of scale is noticed, this source could be rejected. Similarly, the quantity of volcanism (900%) needed would have left surely other traces of geochemical (Dickens et al., 1995). The impact of carbonaceous extraterrestrial bollide is an interesting substitute, having a piece of seldom evidence yet (Kent et al., 2003).

The most significant theory to date for the origin of this concept, carbon appears from the extensive dissociation of methane out of gas hydrates segregated on the continental shelf (Dickens, 2001). Methanogenic bacteria produce methane which devours the organic material in sediment is largely anaerobic cases. Only upon under pressure and cold, methane hydrates (or clathrates) are balanced. The methane gas can be discharged if there is a reduction in the pressure, or the oceans warm. At present, in these hydrates, the carbon has a 60% of an approximate value, and carbon of about 1,500 to 2,000 Gt (1 Gt = 10^{15} g) would be needed to correspond to the CIE. This value is quite below the currently known reservoir of gas hydrates in the ocean. In the mid-Paleocene, if the $d^{13}C$ is maximum, then it might indicate that certainly more methane has been sequestered with the late Paleocene. The additions of carbon to the global ocean are implied by a hypothesis of methane hydrate that is rational now with the extensive carbonate dissolution.

The primary problem of this hypothesis is to decide that how the separation has to be provoked. Either it is induced by a prolonged ocean warming that exceeds a threshold, a rapid ocean circulation changes which is somehow associated with a turn to the formation of WSBW, or a larger slope failure (Bice and Marotzke, 2002; Kennett and Stott, 1991; Katz et al., 1999)? Is the greenhouse effect of the CH_4 (also its oxidation product CO_2) actually causes the warming? All such possibilities have been controlled in many studies of modeling relevant to the PETM.

4.10.4. Modeling

PETM has occurred due to the pecking order of models from the carbon cycle basic box models to full common circulation models of the atmosphere and ocean. To bind the whole amount of depleted carbon, Paleocene carbon cycle box models are applied which usually requires the CIE to be equal to around 1,500–2,000 Gt of carbon (Dickens, 2001). To evaluate the actions, such models are utilized. These models are helpful to eliminate the extra carbon from the systems by increased weathering, increased organic carbon burial, or by recharge of the 'capacitor' made up of methane. Since possibilities are possible that is why not a single technique out of these is removed so far.

There would have a significant effect of inputs by methane on atmospheric chemistry when there is a direct emission into the atmosphere, resulting in raised lifetimes for methane, and a minimized oxidation capacity, CO, and other hydrocarbons (like isoprene). There is a significant radiative effect of such elevated methane (and afterwards CO_2) concentrations and have been measured to be enough to have been the matter of the contemporaneous warming. Nevertheless, uncertainties in the state of background climate, timing, and Paleocene background emissions let this a tentative result at best. However, for polar stratospheric clouds, the somehow raised radiative role is more speculative (because of the raised stratospheric water vapor connected to maximum methane oxidation within the stratosphere).

The usage of complete GCMs has certainly been restricted by the deficiency of well-forced boundary conditions (continental configurations, vegetation distributions, SSTs, etc.); but a sequence of compelling experiments may have associations for the publication of PETM. The insights of the climatic results of such changes might be given by the simulations with the help of the measurements of the differing equator-pole temperature gradient being a forcing function (Huber and Sloan, 2001). But it is not crucial to have persistent with such reconstructions by the models that try to estimate the changes of SST (for example, through maximized GHGs). Specifically, whether the principal westerly wind belts bolster or not over the PETM is dependent on the model (Schmidt, 1999). The results of the ocean-only model which is restricted by the deficiency of an interactive atmosphere, have been utilized to: (a) either radiatively-forced climate alterations are enough to correspond to the alterations in ocean formation reflected in the planktonic and benthic isotope data, (b) predicts either hydrates could be destabilized in enough quantities through a deep-water circulation shift, or others (Bice and Marotzke, 2001, 2002). Yet, no definitive consequences are drawn, and still, there is no study that is able to expose the presence of WSBW. For ocean models, specific dilemmas are determining consistent heat fluxes and surface freshwater, and the depth of significant sills (for example, in the Isthmus of Panama and the Walvis Ridge, and the North Atlantic). Alternate ubiquitous challenges related to GCM simulations of warm climates are discussed here also (for example, seemingly too high tropical warming and too cold continental interiors).

Forthcoming researches are probable to produce more insights into feasible techniques because dynamical oceans are coupled interactively to atmospheric models for this time, and the methane constituent (inclusion of afterward radiative forcing and atmospheric chemistry) is started to take by the carbon cycle models into account. A little study is also needed for the better definition of the vegetation distribution (which has an impact on the chemical emissions, the albedo, and deposition of alternate chemical species (like O_3) which affect the capacity of oxidation of the atmosphere (Beerling, 2000)) and wetland extent and river drainage systems as well.

4.10.5. Mesozoic Analogs

A Hypothesis says that alike CIE have appeared in history, whereas, no other tours have a documented size and quickness of the PETM in the Cenozoic period. However, the similarities are found in the tours related to ocean anoxic events in the Jurassic, Cretaceous, or perhaps at the last of the Neoproterozoic glaciations. Nevertheless, these thinkable analogs will maybe remain vague for some time due to the significantly low quality of data of these previous events. (Hesselbo et al., 2000; Jahren et al., 2001).

REFERENCES

1. Arbic, B. K., Macayeal, D. R., Mitrovica, J. X., & Milne, G. A., (2004). Paleoclimate-Ocean tides and Heinrich events. *Nature, 432*, 460–460.
2. Aubry, M.-P., Berggren, W. A., Stott, L., & Sinha, A., (1996). The upper Paleocene-lower Eocene stratigraphic record and the Paleocene-Eocene boundary carbon isotope excursion: Implications for geochronology. In: Knox, R. W.O., Corfield, R. M., & Dunay, R. E., (eds.), *Correlation of the Early Paleogene in Northwest Europe* (Vol. 101, pp. 353–380). London: Geological Society, Special Publication.
3. Aubry, M.-P., Lucas, S., & Berggren, W. A., (1998). *Late Paleoceneearly Eocene Biotic and Climatic Events in the Marine and Terrestrial Records*. New York: Columbia University Press.
4. Bains, S., Corfield, R. M., & Norris, R. D., (1999). Mechanisms of climate warming at the end of the Paleocene. *Science, 285*, 724–727.
5. Barron, E. J., & Peterson, W. H., (1989). Model simulation of the cretaceous ocean circulation. *Science, 244*, 684–686.
6. Barron, E. J., & Peterson, W. H., (1991). The Cenozoic Ocean circulation based on ocean general circulation model results. *Paleogeogr. Paleoclimatol. Paleoecol., 83*, 1–28.
7. Beerling, D., (2000). Increased terrestrial carbon storage across the Paleocene-Eocene boundary. Paleogeogr. *Paleoclimatol. Paleoecol., 161*, 395–405.
8. Berner, R. A., (1994). 3GEOCARB II: A revised model of atmospheric CO_2 over Phanerozoic time. *Am. J. Sci., 294*, 56–91.
9. Bice, K. L., & Marotzke, J., (2002). Numerical evidence against reversed thermohaline circulation in the warm Paleocene/Eocene Ocean. *J. Geophys. Res., 106*, 11529–11542.
10. Bice, K. L., Barron, E. J., & Peterson, W. H., (1997). Continental runoff and early Cenozoic bottom water sources. *Geology, 25*, 951–954.
11. Bice, K. L., Barron, E. J., Peterson, W. H., (1998). Reconstruction of realistic Early Eocene paleobathymetry and ocean GCM sensitivity to specified ocean basin configuration. In: Crowley, T. J., & Burke, K. C., (eds.), *Tectonic Boundary Conditions for Climate Reconstructions* (Vol. 1, pp. 212–226). New York: Oxford University Press.
12. Bond, G., Kromer, B., Beer, J., Muscheler, R., Evans, M., Showers, W., Hoffmann, S., Lotti-Bond, R., Hajdas, I., & Bonani, G., (2001). Persistent solar influence on North Atlantic surface circulation during the Holocene. *Science, 294*, 2130–2136.

13. Bradley, R. S., (1999). *Paleoclimatology: Reconstructing the Climates of the Quaternary* (2nd edn., Vol. 64, pp. 1–20). International Geophysics, San Diego: Academic Press.
14. Brady, E. C., DeConto, R. M., & Thompson, S. L., (1998). Deep water formation and poleward ocean heat transport in the warm climate extreme of the Cretaceous (80 Ma). *Geophys. Res. Lett., 25*, 4205–4208.
15. Bralower, T. J., Thomas, D. J., Zachos, J. C., Hirschmann, M. M., Röhl, U., Sigurdsson, H., Thomas, E., & Whitney, D. L., (1997). High-resolution records of the late Paleocene thermal maximum and circum-Caribbean volcanism: Is there a link? *Geology, 25*, 963–966.
16. Brazdil, R., Pfister, C., Wanner, H., Storch, H. V., & Luterbacher, J., (2005). Historical climatology in Europe: The state of the art. Clim. *Change, 70*, 363–430.
17. Bush, A. B.G., & Philander, S. G.H., (1997). The late Cretaceous: Simulation with a coupled ocean-atmosphere general circulation model. *Paleoceanography, 12*, 495–516.
18. Bush, A. B.G., (1997). Numerical simulation of the Cretaceous Tethys circumglobal current. *Science, 275*, 807–810.
19. Butzin, M., Prange, M., & Lohmann, G., (2005). Radiocarbon simulations for the glacial ocean: The effects of wind stress, Southern Ocean Sea ice and Heinrich events. *Earth Planet. Sci. Lett., 235*(1–2), 45–61.
20. Carter, R. M., Carter, L., & McCave, I. N., (1996). Current controlled sediment deposition from the shelf to the deep ocean: The Cenozoic evolution of circulation through the SW Pacific gateway. *Geol. Rundsch.,* 438–451.
21. Clark, P. U., Pisias, N. G., Stocker, T. F., & Weaver, A. J., (2002). The role of the thermohaline circulation in abrupt climate change. *Nature, 415*, 863–869.
22. Clarke, G. K.C., Lhomme, N., & Marshall, S. J., (2005). Tracer transport in the Greenland ice sheet: Three-dimensional isotopic stratigraphy. *Quaternary Sci. Rev., 24*, 155–171.
23. Claussen, M., Mysak, L. A., Weaver, A. J., Crucifix, M., Fichefet, T., Loutre, M. F., et al., (2002). Earth system models of intermediate complexity: Closing the gap in the spectrum of climate system models. *Clim. Dynam., 18*, 579–586.

24. CLIMAP-Project Members, (1976). The surface of the ice-age earth. *Science, 191*, 1131–1144.
25. Corfield, R., & Norris, R. D., (1996). Deep water circulation in the Paleocene Ocean. In: Knox, R. W.O. B., Corfield, R. M., & Dunay, R. E., (eds.), *Correlation of the Early Paleogene in Northwest Europe* (pp. 443–456). Geological Society Special Publication.
26. Crowley, T. J., & North, G. R., (1991). *Paleoclimatology*. Oxford Monographs on Geology and Geophysics, No. 18. New York: Oxford University Press.
27. Crowley, T. J., (1998). Significance of tectonic boundary conditions for paleoclimate simulations. In: Crowley, T. J., & Burke, K. C., (eds.), *Tectonic Boundary Conditions for Climate Reconstructions* (pp. 212–226). New York: Oxford University Press.
28. Crucifix, M., (2005). Distribution of carbon isotopes in the glacial ocean: A model study. *Paleoceanography, 20*, 1–18, PA4020.
29. Danabasoglu, G., (2004). A comparison of global ocean circulation model solutions obtained with synchronous and accelerated integration methods. *Ocean Model., 7*, 323–341.
30. Danabasoglu, G., McWilliams, J. C., & Large, W. G., (1996). Approach to equilibrium in accelerated global oceanic models. *J. Climate, 9*, 1092–1110.
31. De Boer, A. M., & Nof, D., (2004). The Bering Strait's grip on the northern hemisphere climate. *Deep Sea Res. I, 51*, 1347–1366.
32. Dickens, G. R., (2001). Carbon addition and removal during the late Paleocene thermal maximum: Basic theory with a preliminary treatment of the isotope record at ocean drilling program site 1051, Blake Nose. In: Norris, R. D., Kroon, D., & Klaus, A., (eds.), *Western North Atlantic Paleogene and Cretaceous Paleoceanography* (pp. 293–306). London: Geological Society.
33. Dickens, G. R., O'Neil, J. R., Rea, D. K., & Owen, R. M., (1995). Dissociation of oceanic methane hydrate as a cause of the carbon isotope excursion at the end of the Paleocene. *Paleoceanography, 10*, 965–971.
34. Doney, S. C., Large, W. G., & Bryan, F. O., (1998). Surface ocean fluxes and water-mass transformation rates in the coupled climate system model. *J. Climate, 9*, 1420–1441.

35. Fischer, G., & Wefer, G., (1999). *Use of Proxies in Paleoceanography* (Vol. 1, pp. 1–23). Examples from the South Atlantic. Berlin: Springer.
36. Gent, P. R., Bryan, F. O., Danabasoglu, G., Doney, S. C., Holland, W. R., Large, W. G., & McWilliams, J. C., (1998). The NCAR climate system model global ocean component. *J. Climate, 11*, 1287–1306.
37. Gills, B. G., & Ray, R. D., (1999). Lunar orbital evolution: A synthesis of recent results. *Geophys. Res. Lett., 26*(19), 3045–3048.
38. Handoh, I. C., Bigg, G. R., & Jones, E. J. W., (2003). Evolution of upwelling in the Atlantic Ocean Basin. *Paleogeogr. Paleoclimatol. Paleoecol., 202*, 31–58.
39. Handoh, I. C., Bigg, G. R., Jones, E. J.W., & Inoue, M., (1999). An ocean modeling study of the Cenomanian Atlantic: Equatorial paleo-upwelling, organic-rich sediments, and the consequences for a connection between the proto-North and South Atlantic. *Geophys. Res. Lett., 26*, 223–226.
40. Haney, R. L., (1971). Surface thermal boundary condition for ocean circulation models. *J. Phys. Oceanogr., 1*, 241–248.
41. Haug, G. H., & Tiedemann, R., (1998). Effect of the formation of the Isthmus of Panama on Atlantic Ocean thermohaline circulation. *Nature, 393*, 673–676.
42. Henderson, G. M., (2002). New proxies for paleoclimate. *Earth Planet. Sci. Lett., 203*, 1–13.
43. Hesselbo, S. P., Grocke, D. R., Jenkyns, H. C., Bjerrum, C. J., Farrimond, P., Morgans-Bell, H. S., & Green, O. R., Massive dissociation of gas hydrate during a Jurassic oceanic anoxic event. *Nature, 406*, 392–395.
44. Hotinski, R. M., & Toggweiler, J. R., (2003). Impact of a Tethyan circumglobal passage on ocean heat transport and "equable" climates. *Paleoceanography, 18*(1), 1007.
45. Hotinski, R. M., Bice, K. L., Kump, L. R., Najjar, R. G., & Arthur, M. A., (2001). Ocean stagnation and end-Permian anoxia. *Geology, 29*, 7–10.
46. Huber, M., & Caballero, R., (2003a). Eocene El Niño: Evidence for robust tropical dynamics in the "hothouse." *Science, 299*, 877–881.
47. Huber, M., & Nof, D., (2006). The ocean circulation in the Southern Hemisphere and its climatic impacts in the Eocene. *Paleogeogr. Paleoclim. Palaecol., 231*, 9–28.

48. Huber, M., & Sloan, L. C., (2001). Heat transport, deep waters, and thermal gradients: Coupled simulation of an Eocene greenhouse climate. *Geophys. Res. Lett., 28*, 3481–3484.
49. Huber, M., Brinkhuis, H., Stickley, C. E., Doos, K., Sluijs, A., Warnaar, J., Schellenberg, S. A., & Williams, G. L., (2004). Eocene circulation of the Southern Ocean: Was Antarctica kept warm by subtropical waters. *Paleoceanography, 19*, PA4026.
50. Huber, M., Sloan, L. C., & Shellito, C., (2003b). Early Paleogene oceans and climate: A fully coupled modelling approach using NCAR's CSM. In: Wing, S. L., Gingerich, P. D., Schmitz, B., & Thomas, E., (eds.), *Causes and Consequences of Globally Warm Climates in the Early Paleogene* (Vol. 369, pp. 25–47). Geolog. Soc. Amer. Special Paper.
51. Hudson, T. L., & Magoon, L. B., (2002). Tectonic controls on greenhouse gas flux to the Paleogene atmosphere from the Gulf of Alaska accretionary prism. *Geology, 30*, 547–550.
52. Kagan, B. A., (1997). Earth-Moon tidal evolution: Model results and observational evidence. *Prog. Oceanogr., 40*, 109–124.
53. Kaplan, J. O., Prentice, I. C., Knorr, W., & Valdes, P. J., (2002). Modeling the dynamics of terrestrial carbon storage since the last glacial maximum. *Geophys. Res. Lett., 29*(22), doi: 10.1029/(2002) GL015230.
54. Katz, M. E., Pak, D. K., Dickens, G. R., & Miller, K. G., (1999). The source and fate of massive carbon input during the latest Paleocene thermal maximum. *Science. 286*, 1531–1533.
55. Keeling, C. D., & Whorf, T. P., (2000). The 1800-year oceanic tidal cycle: A possible cause of rapid climate change. *Proc. Natl. Acad. Sci., 97*(8), 3814–3819.
56. Keigwin, L., (1982). Isotopic paleoceanography of the Caribbean and East Pacific: Role of Panama uplift in late Neogene time. *Science, 217*, 350–353.
57. Kennett, J. P., (1977). Cenozoic evolution of Antarctic glaciations, the circum-Antarctic Ocean and their impact on global paleoceanography. *J. Geophys. Res., 82*, 3843–3860.
58. Kennett, J., & Stott, L. D., (1991). Abrupt deep-sea warming, paleoceanographic changes and benthic extinctions at the end of the Paleocene. *Nature, 353*, 319–322.

59. Kent, D. V., Cramer, B. S., Lanci, L., Wang, D., Wright, J. D., & der Voo, R. V., (2003). A case for a comet impact trigger for the Paleocene/Eocene thermal maximum and carbon isotope excursion. *Earth Planet. Sci. Lett., 211*, 13–26.

60. Koch, P. L., Zachos, J. C., & Gingerich, P., (1992). Correlation between isotope records in marine and continental carbon reservoirs near the Paleocene/Eocene boundary. *Nature, 358*, 319–322.

61. Kutzbach, J. E., & Liu, Z., (1997). Response of the African monsoon to orbital forcing and ocean feedbacks in the middle Holocene. *Science, 278*, 440–443.

62. Lawver, L. A., & Gahagan, L. M., (1998). The initiation of the Antarctic circumpolar current and its impact on Cenozoic climate. In: Crowley, T., & Burke, K., (eds.), *Tectonic Boundary Conditions for Climate Model Simulations* (pp. 213–226). New York: Oxford University Press.

63. Lear, C. H., Rosenthal, Y., & Wright, J. D., (2003). The closing of a seaway: Ocean water masses and global climate change. *Earth Planet. Sci. Lett., 210*, 425–436.

64. LeGrande, A. N., Schmidt, G. A., Shindell, D. T., Field, C. V., Miller, R. L., Koch, D. M., Faluvegi, G., & Hoffmann, G., (2006). Consistent simulations of multiple proxy responses to an abrupt climate change event. *Proc. Natl. Acad. Sci., 103*(4), 837–842.

65. Liu, Z. R., Gallimore, G., Kutzbach, J. E., Xu, W., Golubev, Y., Behling, P., & Selin, R., (1999). Coupled modeling of long-term climate change with the equilibrium asynchronous coupling scheme. *Clim. Dyn., 15*, 325–240.

66. Liu, Z., Lewis, W., & Ganopolski, A., (2004). A coordinated acceleration scheme for the simulation of long-term climate evolution. *Clim. Dyn., 22*, 771–781.

67. Meissner, K. J., Schmittner, A., Weaver, A. J., & Adkins, J. F., (2003). The ventilation of the North Atlantic Ocean during the Last Glacial Maximum: A comparison between simulated and observed radiocarbon ages. *Paleoceanography, 18*(2), 1023.

68. Mikolajewicz, U., Maierreimer, E., Crowley, T. J., & Kim, K. Y., (1993). Effect of Drake and Panamanian Gateways on the circulation of an ocean model. *Paleoceanography, 8,* 409–426.

69. Montenegro, A., Eby, M., Kaplan, J. O., Meissner, K. J., & Weaver, A. J., (2006). Carbon storage on exposed continental shelves during the glacial-interglacial transition. *Geophys. Res. Lett., 33*, L08703.

70. Moore, T. C. Jr., Backman, J., Raffi, I., Nigrini, C., Sanfilippo, A., Palike, H., & Lyle, M., (2004). Paleogene tropical Pacific: Clues to circulation, productivity, and plate motion. *Paleoceanography, 19*(3), PA3013.

71. Motoi, T., Chan, W. L., Minobe, S., & Sumata, H., (2005). North Pacific halocline and cold climate induced by Panamanian Gateway closure in a coupled ocean-atmosphere GCM. *Geophys. Res. Lett., 32*, L10618.

72. Mudelsee, M., & Raymo, M. E., (2005). Slow dynamics of the Northern Hemisphere glaciation. *Paleoceanography, 20*, PA4022.

73. Munk, W., Dzieciuch, M., & Jayne, S., (2001). Millennial climate variability: Is there a tidal connection? *J. Clim., 15*(44), 370–385.

74. Najjar, R. G., Nong, G. T., Seidov, D., & Peterson, W. H., (2002). Modeling geographic impacts on early Eocene Ocean temperature. *Geophys. Res. Lett., 29*, doi: 10.1029/(2001)GL014–438.

75. Nilsson, J., Broström, G., & Walin, G., (2003). The thermohaline circulation and vertical mixing: Does weaker density stratification give stronger overturning? *J. Phys. Oceanogr., 33*, 2781–2795.

76. Nof, D., & Van Gorder, S., (2002). Did an open Panama Isthmus correspond to an invasion of Pacific water into the Atlantic? *J. Phys. Oceanogr., 33*, 1324–1336.

77. Nof, D., (2000). Does the wind control the import and export of the South Atlantic? *J. Phys. Oceanogr., 30*, 2650–2667.

78. Nong, G. T., Najjar, R. G., Seidov, D., & Peterson, W. H., (2000). Simulation of ocean temperature change due to the opening of the Drake Passage. *Geophys. Res. Lett., 27*, 2689–2692.

79. Otto-Bliesner, B. L., Brady, E. C., & Shields, C., (2002). Late cretaceous ocean: Coupled simulations with the NCAR CSM. *J. Geophys. Res., 107*, doi: 10.1029/(2001)JD000821.

80. Paul, A., & Schäfer-Neth, C., (2005). How to combine sparse proxy data and coupled climate models. *Quaternary Sci. Rev., 24*, 1095–1107.

81. Poulsen, C. J., Barron, E. J., Arthur, M. A., & Peterson, W. H., (2001). Response of the mid-Cretaceous global oceanic circulation to tectonic and CO2 forcing. *Paleoceanography, 16*, 576–592.

82. Poulsen, C. J., Jacob, R. L., Pierrehumbert, R. T., & Huynh, T. T., (2002). Testing paleogeographic controls on a Neoproterozoic snowball Earth. *Geophys. Res. Lett., 29*, doi: 10.10129/(2001)GL014352.

83. Poulsen, C. J., Seidov, D., Barron, E. J., & Peterson, W. H., (1998). The impact of paleogeographic evolution on the surface oceanic simulation and the marine environment within the mid-Cretaceous Tethys. *Paleoceanography, 13*, 546–559.

84. Poussart, P. F., Weaver, A. J., & Barnes, C. R., (1999). Late Ordovician glaciation under high atmospheric CO_2: A coupled model analysis. *Paleoceanography, 14*, 542–558.

85. Power, S. B., & Kleeman, R., (1993). Multiple equilibria in a global ocean general circulation model. *J. Phys. Oceanogr., 23*, 1670–1681.

86. Power, S. B., & Kleeman, R., (1994). Surface flux parameterization and the response of OGCMs to high latitude freshening. *Tellus, 46*, 86–95.

87. Quinn, W. H., & Neal, V. T., (1987). El Niño over the past four and a half centuries. *J. Geophys. Res., 92*(C13), 14449–14461.

88. Roche, D., Paillard, D., Ganopolski, A., & Hoffman, G., (2004). Oceanic oxygen-18 at the present day and LGM: Equilibrium simulations with a coupled climate model of intermediate complexity. Earth Planet. *Sci. Lett., 218*, 317–330.

89. Saravanan, R., & McWilliams, J. C., (1995). Multiple equilibria, natural variability, and climate transitions in an idealized ocean-atmosphere model. *J. Climate, 10*, 2296–2323.

90. Sausen, R., & Voss, R., (1996). Techniques for asynchronous and periodically synchronous coupling of atmosphere and ocean models. I. General strategy and application to the cyclo-stationary case. *Clim. Dyn., 12*, 595–604.

91. Schmidt, G. A., (1999). Forward modeling of carbonate proxy data from planktonic foraminifera using oxygen isotope tracers in a global ocean model. *Paleoceanography, 14*, 482–497.

92. Seager, R., Kushnir, Y., & Cane, M. A., (1995). On heat flux boundary conditions for ocean models. *J. Phys. Oceanogr., 25*, 3219–3230.

93. Shaffer, G., & Bendtsen, J., (1994). Role of the Bering strait in controlling North Atlantic Ocean circulation and climate. *Nature, 367*, 354–357.

94. Stocker, T. F., & Wright, D. G., (1996). Rapid changes in ocean circulation and atmospheric radiocarbon. *Paleoceanography, 11*(6), 773–795.

95. Thomas, D. J., (2004). Evidence for production of North Pacific deep waters during the early Cenozoic Greenhouse. *Nature, 430*, 65–68.
96. Toggweiler, J. R., & Bjornsson, H., (2000). Drake Passage and paleoclimate. *J. Quatern. Sci., 15*, 319–238.
97. Toggweiler, J. R., & Bjornsson, H., (2000). Drake passage and paleoclimate. *J. Quaternary Sci., 15*, 319–328.
98. Toggweiler, J. R., & Samuels, B., (1995). Effect of drake passage on the global thermohaline circulation. *Deep Sea Res., 42*, 477–500.
99. Weaver, A. J., Eby, M., Wiebe, E. C., Bitz, C. M., Duffy, P. B., Ewen, T. L., Fanning, A. F., Holland, M. M., Mac-Fadyen, A., Matthews, H. D., Meissner, K. J., Saenko, O., Schmittner, A., Wang, H., & Yoshimori, M., (2001). The UVic earth system climate model: Model description, climatology, and applications to past, present, and future climates. *Atmos. Ocean, 4*, 361–428.
100. Werner, M., Mikolajewicz, U., Hoffmann, G., & Heimann, M., (2000). Possible changes of $d^{18}O$ in precipitation caused by a meltwater event in the North Atlantic. *J. Geophys. Res., 105*(D8), 10161–10168.
101. Williams, G. E., (2000). Geological constraints on the Precambrian history of earth's rotation and the moon's orbit. *Rev. Geophys., 38*(1), 37–59.

Chapter 5

Pliocene Climates

CONTENTS

5.1. Introduction 148
5.2. Climate Dynamics and Climate Modeling 150
5.3. Causes of Pliocene Climate Change Climatic Warmth 156
References 159

5.1. INTRODUCTION

The topmost part of the Tertiary period is represented by the Pliocene epoch. According to Berggren et al. (1995), its geographical time duration is from 5.3 to 1.8 million years in the past. The timespan in which Earth faced a conversion of climate from warmer to cooler in the Quaternary period is included in the epoch.

Based on climate, there may be three stages of Pliocene (a) The warm phase of Pliocene in the beginning, (b) In the middle of the Pliocene (centered almost 3 million years ago), a comparatively shorter period of 'warm blip,' and (c) a during the late Pliocene, the worsening of climate. It is important to understand that the continuous variation in temperature and precipitation helps in the estimation of small subdivisions of the epoch happening on the timescale of thousands of years and this is proposed by existing high resolution climate histories for this time (e.g., Leroy and Dupont, 1994). It is difficult to quantitatively represent climate changes (Haywood et al., 2002a) but usually, climate changes happened with predicted cycles of 41,000 and 19,000, 21,000 and 23,000 years associated with precessional Milankovitch cycles and obliquity (the angle of Earth's axial tilt measured with respect to the plane of the ecliptic; Milankovitch cycles is the cyclic fluctuations of the Earth's orbital position; precession is the degree of the wobble of the Earth on its axis, due to the gravitational pull of the Moon and the Sun. It is possible that the absence of long paleoclimatic information and high resolution makes various climatic events unnoticed which are mostly of short and abrupt timescale, hence procedure for generalizing the climate of the Pliocene becomes tremendously dangerous.

Most of the part of the Pliocene world is warmer as compared to the current situation although there was a continuous increase in cooling during the Tertiary periods. There were usually several degrees higher (3°C to 5°C) mean yearly temperatures in mid-latitudes Tertiary periods than at present and it is specified by an antiquated allocation of marine (fish) microfossil (tiny oceanic creatures having (without diversity of function) single cell or a colonially combined cells) and planktonic (impartial maritime animals) together with terrestrial flora and fossil fauna (e.g., Thompson, 1991; Dowsett et al., 1996). Typically, the temperatures at high latitudes and polar regions were sufficiently hot to permit the class of plants and animals to be at higher latitudes than their adjacent current families and these are the places where utmost warming has been seen (Adam, 1994).

The performance and amount of sea ice for Pliocene Sea levels and the main terrestrial ice sheets of Greenland and Antarctica had the effect of higher temperature at poles and latitudes.

5.1.1. Sea Ice

Data of diatom (see Diatoms) taken from the foraminifer and Southern Ocean (Barron, 1996), mollusk, and ostracod (Ostracodes) data (Cronin et al., 1993) recommends that, with present summer sea-ice limits alike to Pliocene winter sea-ice cover, the amount of sea ice were largely decreased at about 3 million years ago, and it is likely that the Arctic Ocean was occasional free of sea ice (Cronin et al., 1993).

5.1.2. Ice Sheets

An Important discussion on the nature of the terrestrial ice sheets of Antarctica and Greenland throughout the Pliocene has been carried out. The important Northern Hemisphere glaciation started the end of the Pliocene is the consent point of view through the analysis of ice-rafted detritus (IRD) in marine sediment cores (e.g., Kleiven et al., 2002). Though, with current proof for glacial action on Greenland as far away as the late Miocene, there may have been many wrong initiates to this glaciation. Even at the last of the Pliocene epoch, the entire mass of ice on Greenland may have kept crucially less than now.

Two opposing aspects occur according to Antarctica. The first one recommends that steady polar conditions have continued for no less than the previous 14 million years and it is based on geomorphic examinations of the landscape in the Dry Valleys region (Denton et al., 1993). The second one recommends that there may have a large reduction in size (by a maximum of one-third) of the East Antarctic ice sheet (EAIS) and it is suggested based on the proof resulting from the remains of Nothofagus (Southern Beech) in the Sirius Group glacial sediments and the Transantarctic Mountain Range along Pliocene marine diatoms (Harwood and Webb, 1986). The dating of a diversity of landforms that span the Miocene and Pliocene involving old desert pavements, volcanic ash layers, superficial till sheets and avalanche deposits seemingly declare to incredibly slow erosion and weathering rates of weathering in the Dry Valleys (Marchant et al., 1993). It is possible that such geomorphic characteristic, if Miocene or Pliocene actually in age, would have survived, if the EAIS acted in an active way.

5.1.3. Sea Level

In the past worldwide sea levels should have been higher if terrestrial ice cover was decreased in the time of the Pliocene. Though the possible faults in the data make it difficult to measure the actual variation of sea level, but the sign occurs which shows that the sea levels were higher. On the basis of classified stratigraphic methods, Haq et al. (1987) stated the sea-level high stand of up to 60 m.

The micropaleontological sign along with the occurrence of the Orangeburg Scarp (a widespread paleo-shoreline from the Atlantic Coastal Plain in the southeastern U.S.) was used by Dowsett and Cronin (1990) to propose that in past between 3.5 and 3.0 million years ago a sea-level high stand of approximately 3,518 m higher than today.

According to the stratigraphic supply of disconformities (type of unconformity where the suppressed erosion surface lies between two sequences of strata that are similar on a broad scale) at Enewetak Atoll, a sea level of Pliocene is 36 m higher than present (Wardlaw and Quinn, 1991). Sea levels have been estimated as 25 m and 30–35 m higher around 3.0 million years and between 4.0 to 3.5 million years before present respectively than present through the correlation and analysis of $d^{18}O$ isotopic histories from deep-sea centers with Atlantic Coastal Plain sediments (Krantz, 1991). Glacio-eustatic events have determined these sea-level variations but there is vague relation between these events and sea level changes (Dowsett et al., 1996).

5.2. CLIMATE DYNAMICS AND CLIMATE MODELING

Conveniently, for the examination of past climate dynamics, the use of general circulation models (GCMs) or numerical climate models is one of the most powerful tools. With the use of different models of variable complexity, many climates modeling studies have been carried out for the Pliocene (Chandler et al., 1994; Sloan et al., 1996).

The duration of warming, which happened around 3 million years ago, has been the point of focus of these modeling studies. The means of geological and boundary condition data sets (required for the beginning of the climate model experiments) that happen for this period are the main reason for it (Figure 5.1) (Dowsett et al., 1999).

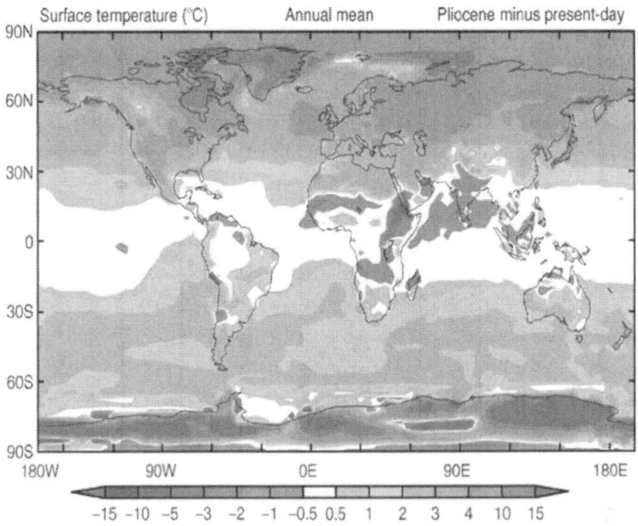

Figure 5.1. Climate model estimation: As a yearly average, the difference between today's surface temperature (C) and simulated Pliocene (3 million years ago) (Haywood et al., 2000a; Haywood, 2001).

Source: https://www.sciencedirect.com/science/article/abs/pii/S0921818100000028X.

Note: The boundary conditions specified from the PRISM2 data set, Hadley Center Atmospheric Climate Model Version 3 (HadAM3) model was used (Dowsett et al., 1999).

5.2.1. Surface Temperature (C)

According to the atmosphere-only climate model studies (experiments in which an interactive ocean model in any type do not involve), the yearly average of worldwide climate around 3 million years ago was 2°C to 5°C warmer than today's (Sloan et al., 1996; Haywood et al., 2000). There are three phases of temperature variation which were estimated in the climate modeling study of Haywood et al. (2000). The 1°C to 4°C was the fair warming estimation between 10 and 30 South and North of the equator. Temperatures warmed by 4–8°C in the Northern Hemisphere between 30 and 50 (Figure 5.2).

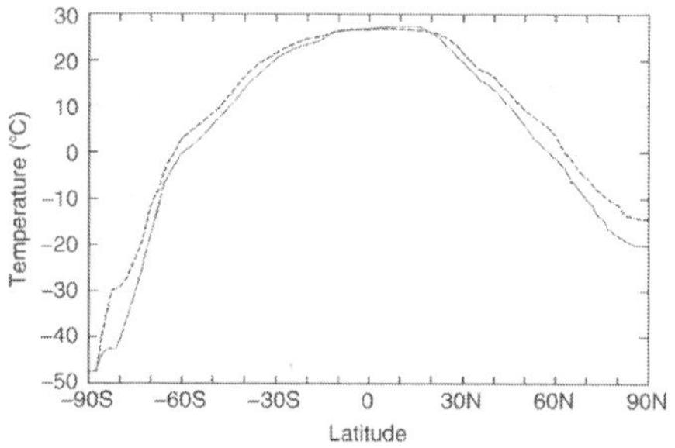

Figure 5.2. As a function of latitude, average yearly surface temperatures (C) plotted for present-day (solid line) and model-predicted Pliocene (dashed line). This plot shows a surface temperature region with larger equability and decreased seasonality at the time of the Pliocene relative to modern (Haywood, 2001).

The temperature increment was 0–4°C in the Southern Hemisphere and this pattern was not repeated there. Significantly in both hemispheres, the highest comparative warming at latitudes was specified as above 70 (Figure 5.2). In high latitude zones of the Northern Hemisphere, the temperature variation as compared to today's fluctuated from 16–20°C in Greenland to 20–28°C in the Arctic in wintertime (December, January, and February; DJF) order. In subdivisions of east Antarctica and west (~32–36°C), in DJF (Southern Hemisphere summer) extreme warming was estimated. Antarctica warmed in a similar pattern to the DJF duration (20–32°C) in June, July, and August (JJA). While, over a large area of the landform, the warming took place (Figure 5.3).

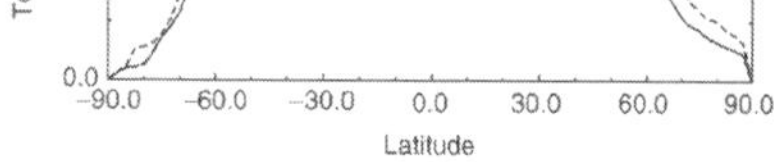

Figure 5.3. As a function of latitude, yearly average total precipitation rate for Pliocene in mm day^{-1} and the present-day (Haywood et al., 2000a; Haywood, 2001).

Source: https://link.springer.com/referenceworkentry/10.1007%2F978-1-4020-4411-3_172.

There was no constant in the pattern of temperature variation. The 10 South to 10 North of the equator expanded over areas of South America and northern Australia are the areas where irregular cooling (maximum of 4°C) was estimated in DJF. In JJA, particularly over areas of southern Asia and Africa, the low latitude /equatorial cooling band developed more uniformly. Relative to the present at 10 North of the equator in areas of Africa, temperatures cooled by 4 and 8°C. The decrease in ice cover in both sea and earth settings during the Pliocene can be due to the estimated high latitude warming (>70 South and North) and it is also the reason for given warmer sea surface temperatures (SSTs) of high latitude. Moreover, it would be feasible not to undervalue the importance of a higher Pliocene Sea level. Higher sea levels particularly in the high latitudes, act as a positive response method for temperature because when shown to cold environments, land cool more than the sea (Sloan et al., 1996). There is more than one aspect contributing to the low latitude and equatorial cooling. There are fluctuations in plants and cloud cover in these parts so for the oceans in low latitudes and equatorial, SST values are expected to have an impact. As a cooling response, larger evaporation is supported by increased precipitation in this area. This is likely the major reason for the tropical cooling experienced in Africa.

5.2.2. Total Precipitation Rate (mm day^{-1})

There is a large similarity between the modern distribution of total precipitation and the one that was simulated by Haywood et al. (2000); and Haywood (2001) during Pliocene (Figure 5.3). At the time of DJF and JJ, an approximate increment from 0- and 2-mm day^{-1} in both hemispheres was estimated. In JJA sporadic pockets of less rainfall (–2 mm day^{-1}) were common but not important. Yearly averaged precipitation showing a 6% increment as compared to modern values, as at the time of Pliocene it was 34.65 cm year^{-1}, 2 cm higher than today (32.84 cm year^{-1}) (Figure 5.3). Zones of greater precipitation were recorded in the Northern Hemisphere (10–70 North) over DJF (Figure 5.4). The Caribbean, northwestern North America, the North Atlantic, and the Indian Ocean, where precipitation was increased by 2–8 mm day^{-1} are included in these areas. There was a link between the warmer surface temperatures and the increment in mid-latitude precipitation in the Northern Hemisphere.

Over JJA, in a narrow belt at low latitudes and equatorial, there was an increment in precipitation. The increase in precipitation from 2- and 10-mm day^{-1} (Figure 5.4) was represented by Arabia, West, and Central Africa, India, the Caribbean, Latin America, and the West Coast of Central.

The model showed that there is no variation in the position of the southern limit of the Hadley Cell (the heat and momentum in the atmosphere at low-latitudes is transported by these cells) up till now. Though, there was an additional shift of the northern limit by 5 of latitude to North. So, there was an increase in Hadley Cell in the Pliocene, which permitted more precipitation to fall over a greater geographical zone. The decrement in the equator-to-pole temperature slope (by 6°C) caused the variation in Hadley Cell. Numerous model feedbacks observed at the low latitude and equatorial such as succeeding decrease in surface temperatures increased precipitation, and evaporation were described by broadening of the Hadley Cell.

5.2.3. Wind (m s^{-1}) and Surface Pressure (hPa)

Compared to the present time, the surface wind power decreased in the Pliocene experiment as a yearly universal mean. This reaction was steady with the model-predicted decrease in the equator-to-pole temperature slope of 6°C, which reduced the circulation of the atmosphere. In the westerly part, most of the changes in wind dynamics were detected.

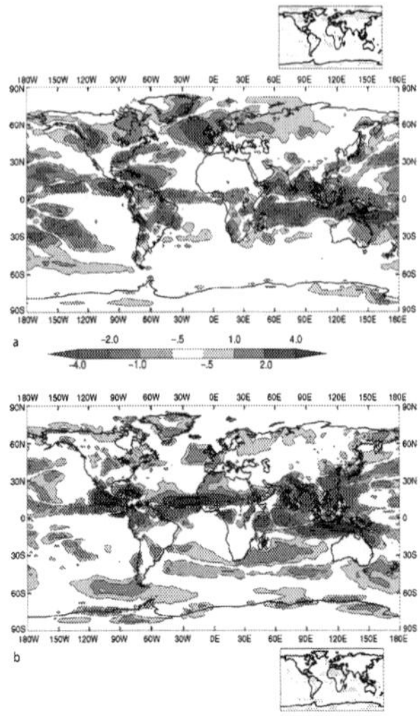

Figure 5.4. Between the Pliocene and current-day for (a) DJF; and (b) JJA, model-predicted alteration in average total precipitation rate (mm day^{-1}). It is observed that the areas where the variation in precipitation is less than 0.5 mm day^{-1} are not shaded and the contour interval is not equally set apart.

Source: https://www.uv.es/~pardomv/pe/2001_1/climate/climate.pdf.

Note: The geographical regions in which the change in precipitation rate is up to a 95% of statistical significance level, are shown by small boxes with grey shading (Haywood et al., 2000; Haywood, 2001).

There is a great similarity of past Pliocene wind directions to the present. Over regions of Siberia, the Tibetan Plateau, and North America, increments in wind speed were detected. This was a function of the changed surface heating and land cover features of the Pliocene simulation relative to the present one. A variation in the Pliocene scenario has a more prominent effect on ocean wind strength. The areas affected by the Aleutian and Icelandic low-pressure systems during DJF have wind speeds enhanced by a maximum of 6 m s^{-1}. Relative to the present these pressure systems expanded in the Pliocene simulation (Figure 5.5).

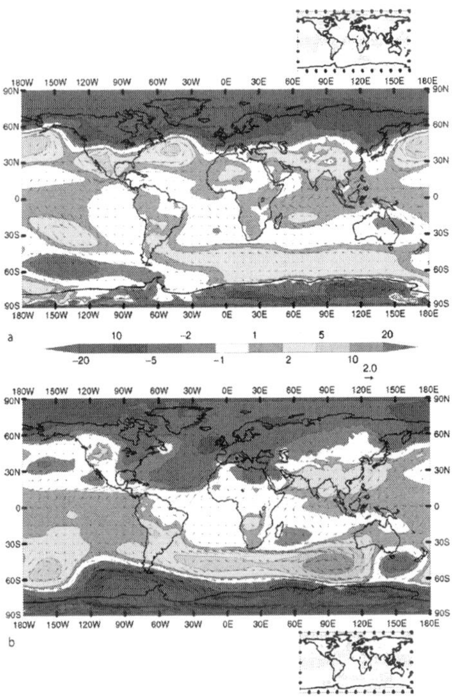

Figure 5.5. Between the Pliocene and current-day for (a) DJF; and (b) JJA, model-predicted alteration in surface pressure (hPa): Vector surface average winds are also presented.

Note: The terrestrial areas in which the difference in precipitation rate is prominent to a 95% of statistical significance level, are presented by mini-boxes with grey shading (Haywood et al., 2000; Haywood, 2001).

At 200 mbs, an investigation of wind strength fields showed that in the Pliocene testing, the intensity of the mid-latitude jet streams was increased. Over subtropic ocean areas, the enhancement in wind speeds in other sites was reported, by increasing high-pressure areas. At the time of JJA and DJF, in the Southern Ocean west of Australia, a powerful high-pressure irregularity was detected (Figure 5.5). In the Pliocene simulation, variations in the Antarctic orography caused Rossby wave (planetary wave) propagation and high-pressure irregularity was linked with the variation in Rossby wave (planetary wave) propagation. In the Pliocene experiment, the intensity of JJA monsoon flow in the Indian Ocean was diminished.

5.3. CAUSES OF PLIOCENE CLIMATE CHANGE CLIMATIC WARMTH

The reasons for Pliocene climate variation have been broadly discussed (Raymo et al., 1996). The question "what was the reason for Pliocene warmth?" is logically arguable and applicable. It is due to (a) equifinality (the formation of the same reaction in climate due to different climate imposing mechanisms), (b) climatic memory effects (the existing universal climate pattern at any one time due to those which happened before as well as imposing mechanisms in process at that pattern), and (c) numerous connections of different parts of the Earth's climate system). "What were the methods that assisted endure, endorse, and finally terminate the climates of the Pliocene?" might be a more logical question. As compared to causation driven, such method has the benefit of operating as to better understand future climate variation, it enhances the advantage of studying past warm durations in our search.

5.3.1. Atmospheric CO_2

Examination of $d^{13}C$ ratios of aquatic organic carbon estimation of the differences between the carbon isotope configuration of surface and deep waters and the investigation of the stomatal density of fossil leaves (a leaf surface has small holes called stomata through which oxygen evolves and carbon dioxide enter the leaf, thus help in photosynthesis; Kürschner et al. (1996) have predicted the atmospheric CO_2 levels (Shackleton et al., 1992). By using these substitution methods, it is recommended that, relative to present concentrations of 375 ppmv and mid-19th century levels of about 280 ppmv, the total CO_2 levels today range from 360 to 400 ppmv. The reason for this higher Pliocene concentration is unclear and may be involved in the change of CO_2 level during Tertiary (Berner, 1990). Otherwise, between many reservoirs (e.g., the land, atmosphere, marine biota, and the oceans, on sediments, rocks, and geological time scales, etc.), the responses between the carbon cycle (the carbon cycling in the form of organic compounds, carbon dioxide, carbonates) and climate may also have importance (Dowsett et al., 1996).

5.3.2. Ocean Heat Transport (OHT)

A variation in the heat transfer of the oceans is possibly an additional feature of consideration. This might be obtained through an increment in surface gyre circulation or thermohaline. In initial modeling studies, this opportunity was underlined but many unsolved problems and difficulties remain unsettled (Rind and Chandler, 1991). Absurdly, a decrease in the slope of latitudinal SST diminishes atmospheric forcing of surface marine motion, and therefore less marine heat transfer takes place from the equator to higher latitudes takes. This is a general issue related to previous warm climate dynamics. Difficulties like (a) problem to produce the correct recreated hemispheric temperature circulation and, (b) problem to assign a thermohaline coupling reason to all ocean basins, occur in any description of strength-based thermohaline circulation of Pliocene warmth (Figure 5.6) (Crowley, 1996).

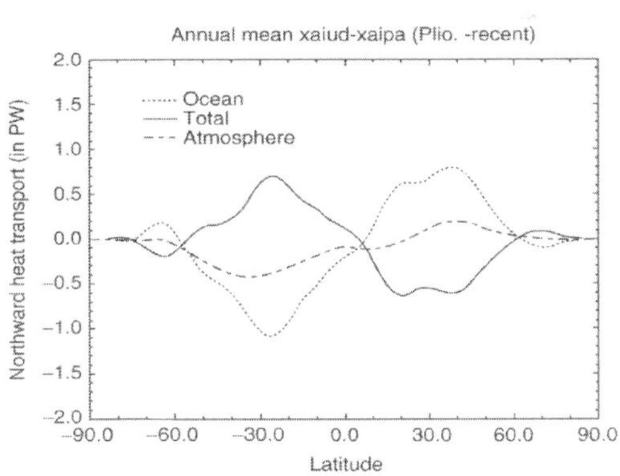

Figure 5.6. Model estimation of the difference in entire poleward heat transfer (in PW), by oceans and atmosphere between the modern-day and the Pliocene climate modeling experiment of (Haywood et al. (2000a) (xajpa = present-day simulation; xaiud = mid-Pliocene simulation).

Source: https://link.springer.com/referenceworkentry/10.1007%2F978-1-4020-4411-3_164.

On the basis of universal scale energetics, above stated arguments are considered. The significance of particular regional features, like fluctuation in wind pressure and surface wind power over different ocean basins, has only been thoroughly considered. Enhancement in westerly constituent winds at middle to high latitudes in both hemispheres is estimated by the outcomes of climates modeling (shown in Figure 5.5). This is feedback to warmer surface temperatures and fewer ice cover over Southern Ocean, Greenland, Antarctica, and the Arctic and to warmer sea-surface temperatures in the South and North Atlantic Oceans on Pliocene. Likewise, in both hemispheres, the increase in sub-tropical high-pressure systems is also observed. Thus, at the time of the Pliocene, regional pressure slopes of both hemispheres in the middle to high latitudes may have been intensified. When linked with warmer SSTs, this can increase the atmospheric transfer of moisture and heat to the high latitudes (Figure 5.6). There is an intensified wind stress (based on modeling experiments, maximum 20 N m^{-2}) to the surface of the oceans due to increment in westerly wind strength, therefore intensifying the flow of heat-transferring surface ocean flows such as the Gulf Stream (Billups et al., 1998).

On the basis of modeling studies, these explanations for improved oceanic heat transfers support paleoceanographical proof (Dowsett et al., 1996). Over Pliocene, uplifting from the California margin and greater carbonate accumulation combines to increase wind pressure (Ravelo et al., 1997). Outcomes from a study of marine ostracodes propose that: (a) During the Pliocene, increased uplifting determined by the Gulf Stream driven by the MidAtlantic coast of the USA and reconstructed cooler ocean temperatures off the southeastern, and (b) North Atlantic circulation may have been strengthened.

REFERENCES

1. Adam, D. P., (1994). Pliocene pollen data set dynamics: Tulelake, California and lost chicken mine, Alaska. In: Thompson, R. S., (ed.), *Pliocene Terrestrial Environments and Data/Model Comparisons* (pp. 6–10). U. S. Geological Survey Open-File Report 94–23.
2. Barron, J. A., (1996). Diatom constraints on the position of the Antarctic Polar Front in the middle part of the Pliocene. *Mar. Micropaleontol., 27*, 195–213.
3. Berggren, W. A., Kent, D. V., Swisher, C. C., & Aubry, M. P., (1995). A revised Cenozoic geochronology and chronostratigraphy. In: Berggren, W. A., Kent, Aubry, M. P., & Hardenbol, J. (eds.), *Geochronology, Time Scales and Global Stratigraphic Correlation* (pp. 129–212). Tulsa, OK: Society for Sedimentary Geology. Special Publication 54.
4. Berner, R. A., (1990). Atmospheric carbon dioxide over Phanerozoic time. *Science, 249*, 1382–1386.
5. Berner, R. A., (1994). Geocarb II: A revised model of atmospheric CO_2 over the Phanerozoic time. *Am. J. Sci., 294*, 56–91.
6. Billups, K., Ravelo, A. C., Zachos, J. C., (1998). Early Pliocene climate: A perspective from the western equatorial Atlantic warm pool. *Paleoceanography, 13*(5), 459–470.
7. Cane, M. A., & Molnar, P., (2001). Closing of the Indonesian sea-way as a precursor to east African aridification around 3–4 million years ago. *Nature, 411*, 157–162.
8. Chandler, M., Rind, D., & Thompson, R., (1994). Joint investigations of the middle Pliocene climate II: GISS GCM Northern hemisphere results. *Glob. Planet. Change, 9*, 197–219.
9. Copeland, P., Harrison, T. M., Kidd, W. S.F., Ronghua, X., & Yuquan, Z., (1987). Rapid early Miocene acceleration of uplift of the Gagdese Belt, Xizang (southern Tibet), and its bearing on accommodation mechanisms of the India-Asia collision. *Earth Planet. Sci. Lett., 86*, 240–252.
10. Cronin, T. M., (1991). Pliocene shallow water paleoceanography of the North Atlantic Ocean based on marine ostracodes. *Quaternary Sci. Rev., 10*(2/3), 175–188.
11. Cronin, T. M., Whatley, R. C., Wood, A., Tsukagoshi, A., Ikeya, N., Brouwers, E. M., & Briggs, W. M., (1993). Microfaunal evidence for elevated mid-Pliocene temperatures in the Arctic Ocean.

Paleoceanography, 8, 161–173.

12. Crowley, T. J., (1991). Modeling Pliocene warmth. *Quaternary Sci. Rev., 10*, 272–282.
13. Crowley, T. J., (1996). Pliocene climates: The nature of the problem. *Mar. Micropaleontol., 27*(1–4), 3–12.
14. Denton, G. H., Sugden, D. E., Marchant, D. R., Hall, B. L., & Wilch, T. I., (1993). East Antarctic ice sheet sensitivity to Pliocene climatic change from a dry Valleys perspective. *Geografiska Annaler, 75A*, 155–204.
15. Dowsett, H. J., & Cronin, T. M., (1990). High eustatic sea level during the Middle Pliocene: Evidence from the southeastern U. S. Atlantic Coastal Plain. *Geology, 18*, 435–438.
16. Dowsett, H. J., Barron, J. A., Poore, R. Z., Thompson, R. S., Cronin, T. M., Ishman, S. E., & Willard, D. A., (1999). *Middle Pliocene Paleoenvironmental Reconstruction: PRISM2*. USGS open file report 99–535. http://pubs.usgs.gov/openfile/of 99–535 (accessed 15 September 2021).
17. Dowsett, H., Barron, J., & Poore, R., (1996). Middle Pliocene Sea surface temperatures: A global reconstruction. *Mar. Micropaleontol., 27*(1–4), 13–26.
18. Francis, J. E., & Hill, R. S., (1996). Fossil plants from the Pliocene Sirius group, Transantarctic mountains: Evidence for climate from growth rings and fossil leaves. *Palaios, 11*, 389–396.
19. Fyles, J. G., Marincovich, L., Matthews, J. V., & Barrendrest, R., (1991). Unique mollusk finds in the Beaufort formation (Pliocene) Meighen Island, arctic Canada. *Curr. Res. Part B, Geological Survey Paper, 91*, 105–112.
20. Haq, B. H., Hardenbol, J., & Vail, P. R., (1987). The new chronostratigraphic basis of Cenozoic and Mesozoic Sea level cycles. *Cush. Found. Foram. Res., Special Publication, 24*, 7–13.
21. Harwood, D. M., & Webb, P. N., (1986). Recycled marine microfossils from basal debris-ice in ice-free valleys of southern Victoria Land. *Antarct. J. US., 21*(5), 87, 88.
22. Haug, G. H., & Tiedemann, R., (1998). Effect of the formation of the Isthmus of Panama on Atlantic Ocean thermohaline circulation. *Nature, 393*, 673–676.

23. Haywood, A. M., & Valdes, P. J., (2004). Modelling Pliocene warmth: Contribution of atmosphere, Oceans and cryosphere. *Earth Planet. Sci. Lett., 218*, 363–377.

24. Haywood, A. M., (2001). *Evaluating General Circulation Climate Model Reliability Against the Pliocene Geological Record* (p. 295). PhD Thesis, Reading University, Reading, Berkshire.

25. Haywood, A. M., Sellwood, B. W., & Valdes, P. J., (2000b). Regional warming: Pliocene (3 Ma) paleoclimate of Europe and the Mediterranean. *Geology, 28*(12), 1063–1066.

26. Haywood, A. M., Valdes, P. J., & Sellwood, B. W., (2002a). Magnitude of climate variability during middle Pliocene warmth: A paleoclimate modeling study. *Paleogeogr. Paleoclimatol. Paleoecol., 188*, 1–24.

27. Haywood, A. M., Valdes, P. J., Francis, J. E., & Sellwood, B. W., (2002c). Global middle Pliocene biome reconstruction: A data/model synthesis. *Geochem. Geophys., Geosyst., 3*(1).

28. Haywood, A. M., Valdes, P. J., Sellwood, B. W., & Kaplan, J. O., (2002b). Antarctic climate during the middle Pliocene: Model sensitivity to ice sheet variation. *Paleogeogr. Paleoclimatol. Paleoecol., 182*, 93–115.

29. Haywood, A. M., Valdes, P. J., Sellwood, B. W., (2000a). Global scale paleoclimate reconstruction of the middle Pliocene climate using the UKMO GCM: Initial results. *Glob. Planet. Change, 25*(1–3), 239–256.

30. Haywood, A. M., Valdes, P. J., Sellwood, B. W., Kaplan, J. O., & Dowsett, H. J., (2001). Modeling middle Pliocene warm climates environments of the U.S.A. *Paleontologia Electronica, 4*(1), https://www.uv.es/~pardomv/pe/2001_1/climate/climate.pdf (accessed 15 September 2021).

31. Kim, S. J., & Crowley, T. J., (2000). Increased Pliocene north Atlantic deep water: Cause or consequence of Pliocene warming. *Paleoceanography, 15*, 451–455.

32. Kleiven, H. F., Jansem, E., Fronval, T., & Smith, T. M., (2002). Intensification of Northern hemisphere glaciations in the Circum Atlantic region (3.5–2.4 Ma)-Ice-rafted detritus evidence. Paleogeogr. *Paleoclimatol. Paleoecol., 184*(3–4), 213–223.

33. Krantz, D. E., (1991). A chronology of Pliocene Sea level fluctuations: The U.S. middle Atlantic Coastal Plain record. *Quaternary Sci. Rev., 10*, 163–174.

34. Kürschner, WM., Van der Burgh, J., Visscher, H., & Dilcher, D. L., (1996). Oak leaves as biosensors of late Neogene and early Pleistocene paleoatmospheric CO_2 concentrations. *Mar. Micropaleontol., 27*(1–4), 299–312.

35. Leroy, S., & Dupont, L., (1994). Development of vegetation and continental aridity in northwestern Africa during the Late Pliocene: The pollen record of ODP Site 658. *Paleogeogr. Paleoclimatol. Paleoecol., 109*, 295–316.

36. Marchant, D. R., Swisher, C. C. III., Lux, D. R., West, D. P. Jr., & Denton, G. H., (1993). Pliocene paleoclimate and East Antarctic Ice-Sheet history from surficial ash deposits. *Science, 260*, 667–670.

37. Marincovich, L. Jr., & Gladenkov A. Yu., (1999). Evidence for an early opening of the Bering Strait. *Nature, 397*, 149–151.

38. Maslin, M. A., Haug, G. H., Sarnthein, M., Tiedemann, R., Erlenkeuser, H., & Stax, R., (1995). Northwest Pacific site 882: The initiation of northern hemisphere glaciation. Ocean drilling program, College Station, TX. In: *Proceedings of the Ocean Drilling Program Scientific Reports. 145*, 315–329.

39. Mercier, J. I., Armijo, R., Tapponnier, P., Carey-Gailhrdis, E., & Lin, T., (1987). Change from late Tertiary compression to Quaternary extension in Southern Tibet during the India-Asia collision. *Tectonics, 6*, 275–304.

40. Mikolajewicz, U., Maier-Reiner, F., Crowley, T. J., & Kim., K. Y., (1993). Effect of Drake and Panamanian gateways on the circulation of an ocean model. *Paleoceanography, 8*, 409–427.

41. Ravelo, A. C., Lyle, M., Koizumi, I., Caulet, J. P., Fornaciari, E., Hayashida, A., Heider, F., Hood, J., Hovan, S., Janecek, T., Janik, A., Stax, R., & Yamamoto, M., (1997). The ODP Leg167 shipboard scientific party. Pliocene carbonate accumulation along the Californian margin. *Paleoceanography, 12*, 729–741.

42. Raymo, M. E., & Rau, G. H., (1992). Plio-Pleistocene atmospheric CO_2 levels inferred from POM d13C at DSDP Site 607. *EOS, Transactions of the American Geophysical Union, Spring Meeting, 73*.

43. Raymo, M. E., (1991). Geochemical evidence supporting T.C. Chamberlin's theory of glaciation. *Geology, 19*, 344–347.

44. Raymo, M. E., Grant, B., Horowitz, M., & Rau, G. H., (1996). Mid-Pliocene warmth: Stronger greenhouse and stronger conveyor. *Mar. Micropaleontol., 27*(1–4), 313–326.

45. Rind, D., & Chandler, M., (1991). Increased ocean heat transports and warmer climate. *J. Geophys. Res., 96*, 7437–7461.

46. Ruddiman, W. F., & Raymo, M. E., (1988). Northern Hemisphere climate regimes during the past 3 Ma: Possible tectonic connections. *Philos. Trans. R. Soc. Lond. B, 318*, 411–430.

47. Shackleton, N. J., Le, J., Mix, A., & Hall, M. A., (1992). Carbon isotope records from Pacific surface waters and atmospheric carbon dioxide. *Quaternary Sci. Rev., 11*, 387–400.

48. Sloan, L. C., Crowley, T. J., & Pollard, D., (1996). Modeling of middle Pliocene climate with the NCAR GENESIS general circulation model. *Mar. Micropaleontol., 27*(1–4), 51–61.

49. Spicer, R. A., Harris, N. B.W., Widdowson, M., Herman, A. B., Guo, S., Valdes, P. J., Wolfe, J. A., & Kelley, S. P., (2003). Constant elevation of southern Tibet over the past 15 million years. *Nature, 421*, 622–624.

50. Thompson, R. S., (1991). Pliocene environments and climates in the western United States. *Quaternary Sci. Rev., 10*(2/3), 115–132.

51. Wardlaw, B. R., & Quinn, T. M., (1991). The record of Pliocene sea-level changes at Enewetak Atoll. *Quaternary Sci. Rev., 10*, 247–258.

52. White, J. M., Ager, T. A., Adam, D. P., Leopold, E. B., Liu, G., Jette, H., & Schweger, C. E., (1997). An 18-million-year record of vegetation and climate change in northwestern Canada and Alaska: Tectonic and global climatic correlates. *Paleogeogr. Paleoclimatol. Paleoecol., 130*(1–4), 293–306.

53. Wright, J. D., & Miller, K. G., (1996). Control of north Atlantic deep-water circulation by the Greenland Scotland Ridge. *Paleoceanography, 11*, 157–170.

Chapter 6

Paleotemperatures and Proxy Reconstructions

CONTENTS

6.1. Introduction 166

6.2. Methods of Paleotemperature Reconstruction 167

6.3. Other Methods 176

References 177

6.1. INTRODUCTION

Most of the energy used to maintain Earth's global temperature comes through incoming solar radiation of various wavelengths, with just a tiny amount of energy coming through geothermal activities. In space, approximately 28% of whole incoming insolation is reflected through molecules, elements, and clouds in the atmosphere. Around 25% are absorbed by water vapor, clouds, and ozone in the atmosphere, and about 47% are immersed in the Earth's surface.

About 18% of this 47% is reflected space, whereas from latent heating the rest heats the atmosphere, convection, and greenhouse gas (GHG) absorption of infrared radiation. Via the atmosphere, because of this absorption of solar energy, the Earth's mean yearly temperature is 33°C higher than it would be else, and projected global warming created as a result of human-caused GHG emissions is a significant concern.

The temperature of the Earth, on the other hand, is not constant. It is controlled by various internal and external mechanisms that operate on different geographic and temporal scales. Variations in the Sun's energy output, the seasonal and geographic distribution of insolation as an outcome of changes in the Earth's orbital configuration, as well as physical (volcanic emissions, wind-blown dust) and chemical features of the atmosphere, along with tectonic methods varying the oceanic circulation and continental distribution.

The oxygen isotopic account for ocean foraminifera, for example (Figure 6.1), shows a vivid picture of climate cooling in excess of the last 100 million years, owing mostly to the effect of long-term variations in atmospheric composition and continental geometries, ocean basin, and further aspects.

The records of temperature (Instrumental and historical) are too little to explain these and further intricate patterns of the temperature of Earth's fluctuation over diverse durations or act as a starting point for evaluating present global temperature trends and the human effect on climate. The interpretation and reconstruction of ocean temperatures and past atmospheric are thus one of the most frequently explored areas in paleoclimatology. This entry works with overall features of the history of Earth's temperature, primarily atmospheric temperature.

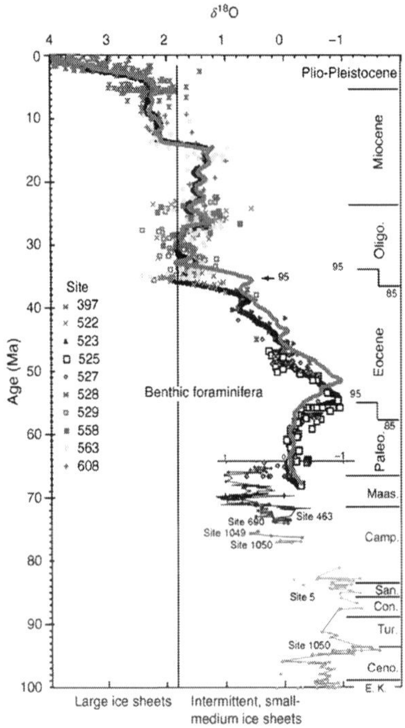

Figure 6.1. Variations in foraminifera's isotopic composition through deep-sea cores show a gradual cooling of the global temperature. Greater negative d18O values characterize times of global warming and low polar ice volume throughout the Cretaceous and Eocene. In contrast, more + values through the Oligocene during Quaternary indicate steady cooling and the formation of polar ice caps.

Source: https://agupubs.onlinelibrary.wiley.com/doi/10.1029/PA002i001p0000 1.

6.2. METHODS OF PALEOTEMPERATURE RECONSTRUCTION

A range of temperature "proxies" that are being used by the paleoclimatologists kept in natural archives like lake sediments, ocean, and tree rings, ice cores (ice sheets) and glaciers, speleothems (precipitated calcite), corals, ancient soils, and further mineralized organisms, and geomorphological and geological characteristics to calculate past temperature changes. Depending on the proxy method employed, paleotemperature reconstructions can be quantitative or qualitative. Nevertheless, all of these methods contain a specific error bar because of shallow knowledge of methods leading to biogeochemical and organic systems, analytical uncertainty, and the effect of features except temperature. Although delays are not enduring, paleotemperature calculations have enhanced markedly in their reliability and accuracy in the last 10 years. A few of the significant uses to reconstruct previous atmospheric temperatures are explained below.

6.2.1. Tree Rings

The research of dendroclimatology, or tree rings (Schweingruber, 1988; Fritts, 1976), depends on the effect that precipitation and temperature have on the development of trees and is very beneficial in the research of climate inconsistency in the past time. Tree rings are made up of units of irregular dark and light layers of latewood and earlywood development that form periodic growth outlines that change from year to year based on the conditions of the environment, the age of the tree, soil, and additional features. Tree ring density, width, and isotopic (hydrogen, oxygen, and carbon) structure of the wood is extensively employed to reconstruct climatic history, particularly in the medium to high latitudes. In boreal trees, summer warmth has a significant impact on latewood density. Dendroclimatology's success is dependent on careful site selection and tree selection, classy numerical analyses to standardize tree ring characteristics to climatic factors, wide cross-calibration of numerous trees for consecutive control, and knowledge of the ecosystem and growth of numerous tree species. For the past few millennia, the widths of the ring have been employed to calculate yearly temperatures in North America (D'Arrigo and Jacoby, 1992). In the example of a 2,000 years temperature record advanced from conifer trees in United States, carbon isotopes provide longer temperature records (Stuiver and Braziunas, 1987).

Recently, researchers in Asia, Europe, and North America have created substantial networks of records of tree ring temperature and further climate indicators like sea level pressure, with more local data from the Southern Hemisphere. Worldwide temperature anomalies and climate changeability and mean annual hemispheric can be predictable for the past 2,000 years through joining the data of tree ring along with further temperature substitution records (such as lakes) and merging the records of proxy with involved records dating back to the 19th century (Mann and Jones, 2003; Figure 6.2). Though the density of Southern Hemisphere sites is insufficient to conclude that region, these studies offer a solid signal that Northern Hemisphere 20th century temperatures are amongst the highest in the past 1,000 years.

They also show how to generate hemispheric and world-scale atmospheric temperature renovations by combining. These and further research also imply that patterns in decadal and interannual climate changeability through the 20th century support the notion that anthropogenetic GHG emissions have altered worldwide climate (Table 6.1).

Table 6.1. Summary of Methods Employed to Reconstruct Atmospheric Temperatures

Paleotemperatures and Proxy Reconstructions			
Paleoclimate Archive[a]	**Proxy Method for Temperature**	**Complications and Comments**	**References**[b]
Speleothems (stalactites, stalagmites, vein calcite)	• Oxygen isotopes	• Effects of temperature on calcite (aragonite) deposition; local precipitation composition	McDermott et al. (2001); Winograd et al. (1992)
Lake peats and sediments	• Insects (primarily beetles) • Fossil pollen • Ostracode shell trace element composition Diatoms • Ostracode shell isotopic composition		Elias and Coope (1994); Prentice et al. (1991); Guiot et al. (1989); Von Grafenstein et al. (1999); Holmes and Chivas (2002)
Ice cores and ice sheets	• Hydrogen and oxygen isotopes • Excess Deuterium (melt layers)	• Changes in seasonal precipitation and moisture source (summer only)	Grootes et al. (1993); Jouzel et al. (1997); Bradley (1999)

Tree rings (dendroclimatology)	• Ring width and ring density • Hydrogen, oxygen, carbon isotopic composition	• Usually less than 1,000 years; standardization because of growth • Precipitation and other features significant; complex biological and hydrological procedures during wood growth	Fritts (1976), Schweingruber (1988); Briffa and Matthews (2002), Mann et al. (1999); Mann and Jones (2003)
Geographic changes, tree lines	• Pollen	• Primarily employed in Arctic region	–
Other treelines, elevation variations	Pollen, macroflora from packrat middens	Discontinuous, mostly glacial-interglacial periods	Betancourt et al. (1990)
Geological and geomorphological features	• Periglacial sediments • Lithology and mineralogical associations • Paleosols (fossil soils) • Glacial sediments	• Rainfall has a sturdy impact. • High latitudes, elevations • Only near glacial margins (qualitative)	Retallak (1990); Washburn (1979); Parrish (1998)
Plant macrofossils	Plant associations, leaf morphology	–	–

[a]*General paleoclimatology texts with the argument of paleotemperature reconstruction.*

[b]These are either complete texts on the proxy technique or essential applications.

6.2.2. Ice Core Studies of Ice Sheets and Glaciers

Since the isotopic composition of water is largely reliant on the air temperature at the rain, water stuck as continental ice sheets and ice in glaciers gives importance of data about the temperature history of high-rise regions and high latitudes. Paleotemperature records dating back 110,000 and 700,000 years were discovered in ice cores from Antarctica and Greenland, respectively.

The oxygen ($^{18}O/^{16}O$ ratio) and hydrogen (deuterium = ^{2}H) isotope composition of the ice is the most extensively utilized paleoclimate proxies improved through ice cores to infer previous air temperatures. Water condensed from vapor is especially deepened in weightier isotopes because of the lower vapor pressure of water molecules bearing heavy isotopes of O (^{18}O) and H (D).

As the temperature drops, the enrichment in condensation decreases. Reliant on climatic parameters and geographic location, the link between $d^{18}O$ of precipitation and the atmospheric temperature has been described in a variety of formulae.

Variations in moisture supply and seasonal variations in precipitation can to affect the oxygen isotopic conformation of the water, complicating ^{18}O-temperature relationship. Over longer timescales, these processes are essential, and researchers must design methods to sort them out to get paleotemperature bends.

The or d, deuterium excess, is a link among the differential fractional process of oxygen and hydrogen during precipitation that can be exploited. The worldwide link between hydrogen and oxygen during precipitation is $dD = 8\ d^{18}O + 10$.

The value + 10 is known as d, reflecting a balance among oxygen and hydrogen isotopes. As atmospheric circumstances change with time, the value for d vagaries as well, and a measure of d through time can be determined by measuring both $d^{18}O$ and dD in ice cores. The impact of kinetic methods through condensation and evaporation and varying moisture sources and water temperature is complicated and must be straightened out before using d to research paleotemperature.

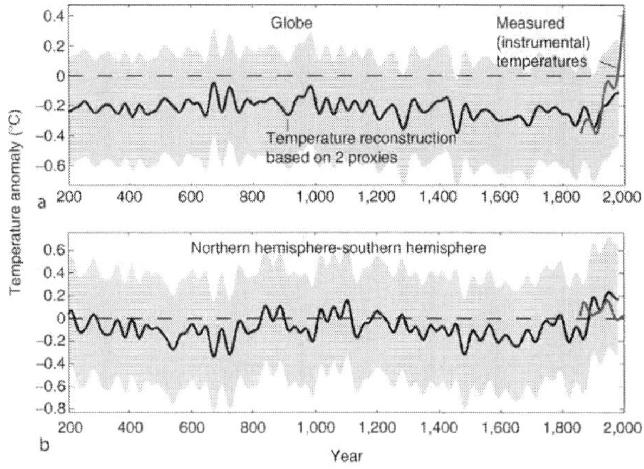

Figure 6.2. For the last 1,800 years, tree rings, lake, and estuarine sediments, and arctic and alpine ice cores have been used to reconstruct and measure global mean temperatures. (a) Reconstructed temperatures combination Southern and Northern Hemisphere records. (b) Centered on the same instrumental and reconstructed temperature curves, hemispheric temperature difference (Mann and Jones, 2003).

Source: https://agupubs.onlinelibrary.wiley.com/doi/abs/10.1029/1999GL900070.

Despite these and other complications, hydrogen, and oxygen isotope records from Antarctica and Greenland have yielded the most significant discoveries regarding Earth's climate past. Two great research projects executed under the sponsorships of the European research program known as GRIP and its American equal,

Greenland Ice Sheet Project II (GISP II), have given outstanding isotopic records of past temperature changeability in the Summit region of Greenland that have been published.

The GRIP oxygen isotope record for the past 100,000 years is depicted in Figure 6.3, revealing severe temperature-related isotopic constancy over the Holocene and the presence of 20 millennial-scale climatic oscillations among 100,000 and 15,000 years ago.

These episodes, known as Dansgaard Oeschger events, represent extraordinary climate uncertainty during the last glacial epoch and are documented in the GISP II core and various paleoclimate records worldwide. The interpretation of the temporal changeability of the Greenland $d^{18}O$-temperature association has enhanced depend on borehole standardization and further means because the Greenland ice core isotopic variations were initially defined and suggests a glacial to an interglacial temperature variance of up to 15°C.

6.2.3. Speleothems and Precipitated Calcite

Vein calcite is known as a low-temperature hydrothermal deposit that forms when groundwater is heated. When calcite or aragonite precipitates from groundwater, it forms stalagmites and stalactites known as speleothems. Because the oxygen fractionation that happens during the deposition of calcitic rock relies on surface temperature, speleothems, and vein calcite can provide crucial paleoclimate information when tested at precise sub-millimeter-scale intervals.

However, because source water precipitation is temperature-dependent and changes because of kinetic processes, understanding of speleothem oxygen isotope records must be done wisely and on basis of case-by-case. From Ireland, significant climate fluctuations in the regional climate in an outstanding centennial-scale speleothem record is being detected. They ascribed to both the impacts of temperature and variations in water vapor causes.

Winograd, Landwehr, and colleagues used vein calcite at Devils Hole in Nevada to provide an essential but contentious late Quaternary paleotemperature record showing primarily western United States and eastern Pacific Sea surface atmospheric temperatures (Figure 6.4). The DH record is remarkable in that, unlike marine records dating back 400,000 years, it is simply dated using uranium sequence approaches, providing evidence for the Milankovitch climate theory.

In its evident design of 100,000-year oddness cycles, the DH temperature record bears a striking resemblance to many other climatic datasets. Figure 6.4, for example, ice core in Antarctica is being linked by the DH isotope record to the deuterium records through the Vostok. Nevertheless, the Devils Hole record specifies that global warming at the start of the latest interglacial era started around 135,000 years ago, some 8,000–10,000 years before the Milankovitch theory of weather expected. The Milankovitch theory faces an unresolved difficulty because of the early age of the penultimate interglacial.

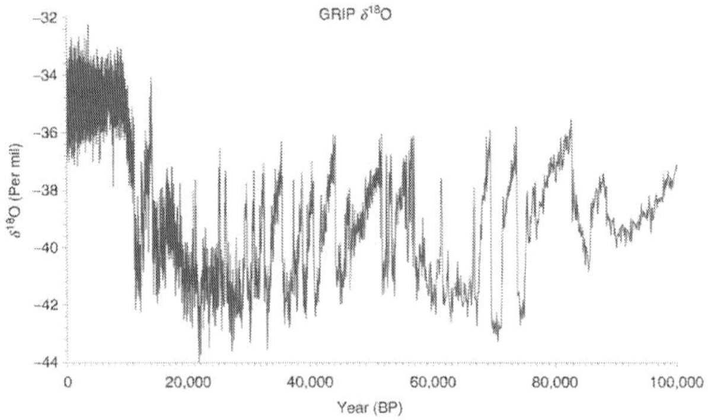

Figure 6.3. For the past 100,000 years, GRIP ice core isotope record Greenland, Summit. On average, the oxygen isotope is 55 cm per mil.

Source: https://www.nature.com/articles/364218a0.

Note: Isotopic excursions among 90,000 and 15,000 yBP signify Dansgaard-Oeschger visionary climate events when fast atmospheric cooling and warming happened over Greenland and several other parts of the world.

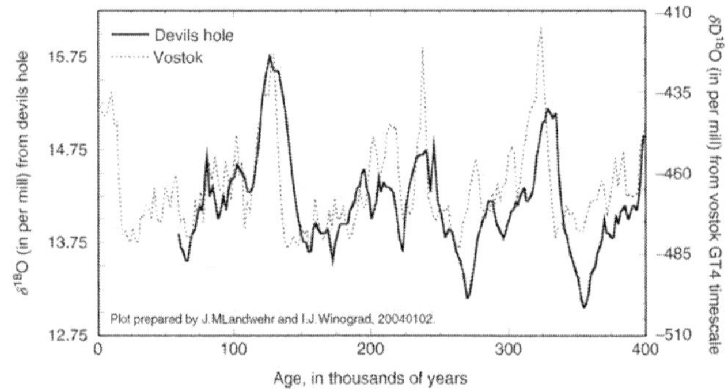

Figure 6.4. For the past 400,000 years, the Nevada, Devils Hole, core DH-11 vein calcite oxygen-18 and the Vostok, Antarctica, ice core deuterium time series.

Source: https://agupubs.onlinelibrary.wiley.com/doi/pdf/10.1029/2001JD900065.

Note: The Devils Hole record shows north eastern Pacific temperatures, while the Vostok deuterium record reflects regional eastern Antarctic temperatures with a distance of 114°.

6.2.4. Lake Records of Temperature

Researchers used various faunal, floral, and geochemical representations to recreate air temperatures through lake sediment cores. Using multiple quantitative statistical approaches, fossil pollen accumulations are often employed to infer previous temperatures. The modern analog technique (MAT) compares ancient accumulations to a vast current pollen database related to summer and winter climate variables (Figure 6.5).

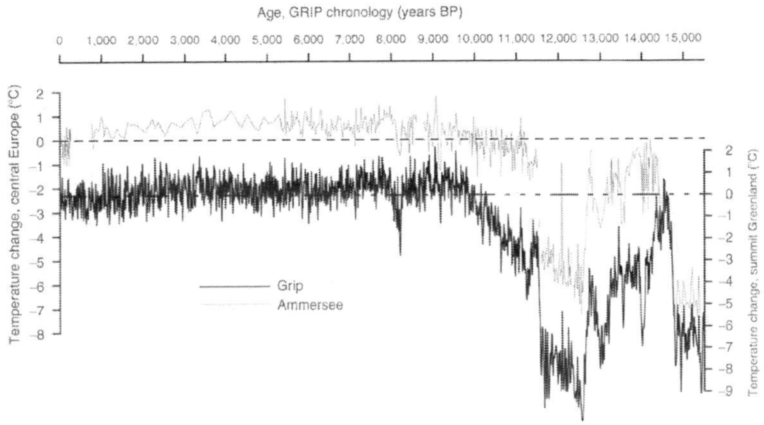

Figure 6.5. Temperature fluctuations at Summit, Greenland GRIP ice core reconstructed through oxygen isotopes, and central European temperatures reconstructed through Ammersee ostracode oxygen isotopes.

Source: https://science.sciencemag.org/content/284/5420/1654#:~:text=Oxygen%2Disotope%20ratios%20of%20precipitation,before%20the%20present%20(B.P.).

Except for periods of rapid climatic change, when pollen and forest accumulations have no current analogs, the MAT approach for North America has produced outstanding temperature history through the past 15,000 years. A 140,000-year temperature record from Les Echets and La Grande Pile in France is being rebuilt by Guiot et al. employing a somewhat different statistical method to fossil pollen. In comparison to today, these analyses demonstrate that in France the mean annual temperature ranged from 12°C to 13°C during peak glacial conditions to 2°C to 4°C during the Eemian interglacial.

Quantitative estimates of atmospheric and aquatic temperatures can also be obtained using geochemical approaches employed to fossil lacustrine ostracodes improved from lake cores. Von Grafenstein et al. (1999) used the $d^{18}O$ of fossil ostracodes from Lake Ammersee, to rebuild a 15,500-year temperature record utilizing the link among mid-European oxygen isotopic configuration of atmospheric and precipitation temperatures (Figure 6.5). The researchers correlated fine-scale European temperature variations throughout the deglacial time with the Greenland record thanks to the decadal resolution reached Ammersee.

6.3. OTHER METHODS

Qualitative (such as., reasonably warm, chilly) and semi-quantitative (i.e., < 4°C, among 15°C and 20°C) clarifications of paleotemperature through the geological record have been used in a variety of ways. Expert explanation of periglacial and glacial deposits and geomorphological features, lithologic forms (e.g., peat, gypsum) and paleogeographic distribution, morphologies, and fossil plant associations, ancient soils, primeval snowline and treeline positions, and other factors are used in these approaches. Few of these signs are impacted through precipitation and other climatic conditions as they are via temperature, as is the case with more quantitative measures. Even when precise temperature estimates are possible, as, in fossil creatures, several of these records are spatially and temporally irregular, making interregional paleotemperature correlations inferring causal variables were challenging. Nonetheless, these methods produce rich and significant literature for many planet locations, giving regional and local information on seasonal temperature or mean annual conditions over many geological periods.

REFERENCES

1. Betancourt, J. L., Van Devender, T. R., & Martin, P. S., (1990). *Packrat Middens: The Last 40,000 Years of Biotic Change* (p. 467). Tucson, AZ: University of Arizona Press.
2. Bradley, R. S., (1999). *Paleoclimatology, Reconstructing Climates of the Quaternary* (2nd edn., p. 613). San Diego, CA: Academic Press.
3. Briffa, K. R., & Matthews, J. A., (2002). Analysis of dendrochronological variability and associated natural climates in Eurasia (ADVANCE10k). *The Holocene, 12*(6), 639–796.
4. Cook, E. R., D'Arrigo, R. D., & Mann, M. E., (2002). A well-verified, proxy reconstruction of the winter North Atlantic Oscillation index since AD 1400. *J. Clim., 15*, 1754–1764.
5. Coope, G. R., (1994). The response of insect faunas to glacial-interglacial climatic fluctuation. *Philos. Trans. R. Soc. Lond., 344B*, 19–26.
6. Cronin, T. M., (1999). *Principles of Paleoclimatology* (p. 560). New York: Columbia University Press.
7. Crowley T. J., & North, G. R., (1991). *Paleoclimatology* (p. 339). New York: Oxford University Press.
8. Cuffey, K. M., Clow, G. D., Alley, R. B., Stuiver, M., Waddington, E. D., & Saltus, R. W., (1995). Large Arctic temperature change at the Wisconsin-Holocene transition. *Science, 270*, 455–458.
9. Curtis, J. H., & Hodell, D. A., (1993). An isotopic and trace element study of ostracodes from Lake Miragone, Haiti: A 10,500-year record of paleosalinity and paleotemperature changes in the Caribbean. In: Swart, P. K., Lohman, K. C., McKenzie, J., & Savin, S., (eds.), *Climate Change in Isotopic Records* (Vol. 78, pp. 135–152). Washington, DC: American Geophysical Union. Geophysical Monograph.
10. D'Arrigo, R. D., & Jacoby, G., (1992). Dendroclimatic evidence from North American. In: Bradley, R. S., & Jones, P. D., (eds.), *Climate Since AD 1500* (pp. 296–311). London: Routledge.
11. Dansgaard, W., Johnsen, S. J., Clausen, H. B., Dahl-Jensen, D., Gundestrup, N. S., Hammer, C. U., Hvidberg, C. S., Steffensen, J. P., Sveinbjornsdottir, A. E., Jouzel, J., & Bond, G., (1993). Evidence for general instability of past climate from a 250-kyr ice-core record. *Nature, 364*, 218–220.

12. Elias, S. A., & Coope, G. R., (1994). *Quaternary Insects and Their Environments* (p. 256). Washington, DC: Smithsonian Institution Press.
13. Frakes, L. A., Francis, J. E., & Syktus, J. I., (1992). *Climate Modes of the Phanerozoic* (p. 286). Cambridge, UK: Cambridge University Press.
14. Fritts, H. C., (1976). *Tree Rings and Climate* (p. 567). London: Academic Press.
15. GRIP Project Members, (1993). Climate instability during the last interglacial period recorded in the GRIP ice core. *Nature, 364*, 203–207.
16. Grootes, P. M., Stuiver, M., White, J. W.C., Johnsen, S., & Jouzel, J., (1993). Comparison of oxygen isotope records from the GISP2 and GRIP Greenland ice cores. *Nature, 366*, 552–554.
17. Guiot, J., Pons, A., de Beaulieu, J.-L., & Reille, M., (1989). A 140,000-year continental climate reconstruction from two European pollen records. *Nature, 338*, 309–313.
18. Holmes, J. A., & Chivas, A R., (2002). *The Ostracoda: Applications in Quaternary Research, Geophysical Monograph Series* (Vol. 131). Washington, DC: American Geophysical Union.
19. Huber, B. T., Norris, R. D., & MacLeod, K. G., (2002). Deep sea paleotemperature record of extreme warmth during the Cretaceous. *Geology, 30*, 123–126.
20. Johnsen, S. J., Dahl-Jensen, D., Dansgaard, W., & Gundestrup, N. S., (1995). Greenland paleotemperatures derived from GRIP borehole temperature and ice core isotope profiles. *Tellus, 47B*, 624–629.
21. Jouzel, J., Alley, R. B., Cuffey, K. M., Dansgaard, W., Grootes, P., Hoffmann, G., Johnsen, S. J., Koster, R. D., Peel, D., Shuman, C. A., Stievenard, M., Stuiver, M., & White, J., (1997). Validity of the temperature reconstruction from water isotopes in ice cores. *J. Geophys. Res., 102*, 26471–26487.
22. Jouzel, J., Lorius, C., Petit, J. R., Genthon, C., Barkov, N. I., Kotlyakov, V. M., & Petrov, V. N., (1987). Vostok ice core: A continuous isotopic temperature record over the last climatic cycle (160,000 years). *Nature, 329*, 403–408.
23. Landwehr, J. M., & Winograd, I. J., (2001). Dating the Vostok ice core record by importing the Devils Hole chronology. *J. Geophys. Res., 106*(D23), 31853–31861.

24. Landwehr, J. M., Coplen, T. B., Ludwig, K. R., Winograd, I. J., & Riggs, A. C., (1997). *Data from Devils Hole core DH-11. U.S. Geol. Survey Open-File Report* 97–79, pp. 2, 8.
25. Mann, M. E., & Jones, P. D., (2003). Global surface temperature over the past two millennia. *Geophys. Res. Lett., 30*(15), 1820.
26. Mann, M. E., Bradley, R. S., & Hughes, M., (1999). Northern Hemisphere temperatures during the millennium: Inferences, uncertainties, and limitations. *Geophys. Res. Lett., 26*, 759–762.
27. McDermott, F., Mattey, D. P., & Hawkesworth, C., (2001). Centennial-scale Holocene climate variability revealed by a high-resolution speleothem $d^{18}O$ record from SW Ireland. *Science, 294*, 1328–1331.
28. Miller, K. G., Fairbanks, R. K., & Mountain, G. S., (1987). Tertiary oxygen isotope synthesis, sea level and continental margin erosion. *Paleoceanography, 2*, 1–19.
29. Parrish, J. T., (1998). *Interpreting Pre-Quaternary Climate from the Geologic Record* (p. 338). New York: Columbia University Press.
30. Petit, J. R., Jouzel, J., Raynaud, D., Barkov, N. I., Barnola, J.-M., Basile, I., Bender, M., Chappellaz, J., Davis, M., Delaygue, G., Delmotte, M., Kotlyakov, V. M., Legrand, M., Lipenkov, V. Y., Lorius, C., Pepin, L., Ritz, C., Saltzman, E., & Stievenard, M., (1999). Climate and atmospheric history of the past 420,000 years from the Vostok ice core, Antarctica. *Nature, 399*, 429–436. Data available from NOAA at: http://www.ngdc.noaa.gov (accessed 15 September 2021).
31. Prentice, I. C., Bartlein, P. J., & Webb, T. W. III., (1991). Vegetation and climate in eastern North American since the last glacial maximum. *Ecology, 72*, 2038–2056.
32. Retallak, G. J., (1990). *Soils of the Past* (p. 512). Boston, MA: Unwin Hyman.
33. Rozanski, K., Araguás, L., & Gofiantini, R., (1993). Isotopic patterns in modern global precipitation. In: Swart, P. K., Lohman, K. C., McKenzie, J., & Savin, S., (eds.), *Climate Change in Continental Isotopic Records* (Vol. 78, pp. 1–36). Washington, DC: American Geophysical Union. Geophysical Monograph.
34. Schweingruber, F. H., (1988). *Tree Rings: Basics and Applications of Dendrochronology* (p. 276). Dordrecht: D. Reidel.
35. Stuiver, M., & Braziunas, T. F., (1987). Tree cellulose $^{13}C/^{12}C$ isotopic ratios and climate change. *Nature, 328*, 58–60.

36. Von Grafenstein, U., Erlenkeuser, H., Brauer, A., Jouzel, J., & Johnsen, S. J., (1999). A mid-European decadal isotope-climate record from 15,500 to 5,000 years BP. *Science, 284*, 1654–1657.
37. Washburn, A. L., (1979). *Geocryology* (p. 406). London: Arnold.
38. Winograd, I. J., Coplen, T. B., Landwehr, J. M., Riggs, A. C., Ludwig, K. R., Szabo, B., Kolesar, P. T., & Revesz, K. M., (1992). Continuous 500,000-year climate record from vein calcite in Devils Hole, Nevada. *Science, 258*, 255–260.
39. Winograd, I. J., Landwehr, J. M., Ludwig, K. R., Coplen, T. B., & Riggs, A. C., (1997). Duration and structure of the past four interglaciations. *Quaternary Res., 48*, 141–154.

Chapter 7

Holocene Climate

CONTENTS

7.1. Introduction 182

7.2. Data Sources 183

7.3. Climate Change During the Latest Deglacial Period 184

7.4. The Climatic Optimum of the Earlier Holocene 185

7.5. Climate Variability in the Late Holocene 187

7.6. Variation of Holocene Treeline 190

References 197

7.1. INTRODUCTION

The Holocene epoch, which started since the last glaciation terminated and is often referred to as the post-glacial interval, is the most recent period of the Planet's history. The Holocene is one of many interglaciations that has occurred during the late Cenozoic glacial period. Gervais coined the term in 1869, and the International Geological Congress recognized it in 1885 as part of viable geological nomenclature. Roberts provides a brief overview of the Holocene environmental record-1998. The Holocene is a focal point of the Past Global Changes-PAGES-program, which has its commission within the International Union for Quaternary Research (INQUA). The Holocene, a journal dedicated solely to Holocene study, has been published since 1991.

In terms of how the Holocene must be officially described, there are three different theories. So, the Holocene is still "alive," discussion has centered on determining when it began, specifically the Pleistocene-Holocene threshold. Wright and Watson (1980) opine the first perspective that the commencement of the Holocene can be effortlessly identified in individual statistics of biotic change-for instance, pollen diagrams-or deglacierization, nevertheless that the Pleistocene Holocene delimitation should be time breach as these alterations are not precisely the same stage all around. The second point of view favors the-type section-method of defining stratigraphic limits, which is frequently used in earth science. In 1976, Mörner for instance, suggested that the Pleistocene-Holocene boundary be defined by sequential order in southern Sweden. The third viewpoint, which is shared by INQUA, is that the Holocene began 10,000 radiocarbons (C) years earlier.

An organized variation between tree-ring ages and ^{14}C and has been discovered in specimens older than 2,500 years in northwest Europe and the western United States, with a 600-year or 12% difference at 5,000 (yBP) years before. Furthermore, the standard curve is not a straight line, but rather a series of deviations that show short-term differences in atmospheric ^{14}C production. Because some ^{14}C dates have multiple calendar ages, this is important. Gulliksen et al. (1998); and Alley et al. (1993) Focused on tree ring counts and yearly ice layers, the latest projections for the onset of the Holocene place it around 11,500 years ago. According to Reimer and Stuiver (1993), it is preferable to use academic years instead of radiocarbon years as a period for previous global changes, so it is now feasible to calibrate ^{14}C schedules for the entire Holocene. In this case, the erstwhile (cal. yBP) will be used.

Attempts to categorize the Holocene have been formed in the past, usually based on derived climatic alteration. For example, premised on transitions in peat stratigraphy in northern Europe, Sernander, and Blytt suggested a method that alternated cool wet and warm-dry cycles. Conversely, orderly ^{14}C dating has revealed that Holocene flora and climatic variations were frequently time-transgressive, besides as a consequence, the majority of researchers favor using age-for instance, 8,000 cal. yBP-instead of the stage-for example, Boreal-Atlantic shift-to define historical events (Figure 7.1).

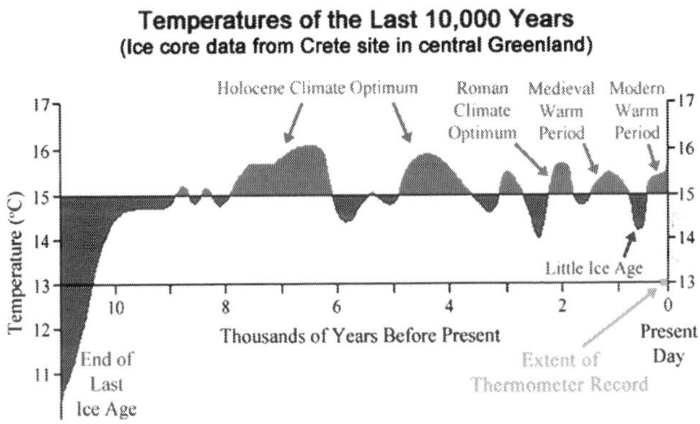

Figure 7.1. Variability in Holocene climate across time.

Source: https://www.researchgate.net/figure/Holocene-climate-variability-over-the-period-of-11000-years_fig1_320191078.

7.2. DATA SOURCES

Instrumental statistics are available over the last few millennia of the Holocene and can be utilized to construct high-grade time series for individual climatic parameters. Besides that, so the climate system changes over longer timeframes than can be directly measured, long-lived natural and anthropogenic records must also be used. According to Bradley and Jones (1992), Documentary sources can be exercised for medium time frames, while these statistics were not always gathered in a format that is ideal for the observation we want to perform. In some regions of the world, the "ancient" time frame can be as long as 5,000 years in Egypt, but only a few hundreds of years in Papua-New Guinea. The 3rd set of techniques works on Holocene time horizons that are longer. According to Webb et al.

(1993), pollen evaluation and dendro-climatology are examples of proxy-climate techniques that can be integrated into a single arrangement to form a multiproxy technique.

7.3. CLIMATE CHANGE DURING THE LATEST DEGLACIAL PERIOD

According to tuned radiocarbon dating, the Last Glacial all-out took place between 25,000 to 18,000 years ago. Temperatures on earth dropped by up to 20°C during this time, while ocean temperatures remained relatively stable. Smaller icebergs persisted in the Alps, parts of East Asia, and the southern Andes during the full icy climate. while there may have been roughly asynchronism between the southern and Northern Hemispheres, climatic warming and subsequent deglacierization of these continental areas occurred in a series of actions that seem to have been extensively synchronous across large parts of the world.

De-glaciation was an erratic method in North America and Europe, taking approximately 9,000 years to finish, and icecaps momentarily up to date at different times towards the termination of the Pleistocene. This was the outcome of a brief but sharp change in climate, which served as an incorrect start to the current interglacial period in at least certain scenarios. This late Pleistocene thermal swaying in the circumference of North Atlantic locality included a temperature rise acknowledged as the Billing-Allerd interstadial (15,000 to 12,700 cal. yBP) accompanied by a cooling event identified as the Younger Dryas stadial-12,700 to 11,500 cal. yBP. Summer temperatures quickly warmed to attain-or even surpass-those of today shortly after 15,000 cal. yBP, according to scientific proof from beetle and midge remnant.

Ice cores contain some of the most crucial paleoclimatic statistics for glacial period climate, with Greenland's database being the most vital in the North Pole. Because ice accumulates more quickly here than in Antarctica, it provides a more detailed record of past climate changes, with a yearly time resolution for the late Pleistocene and Holocene periods. The cores-GISP2 and GRIP, which were discussed at the summit meeting of the Greenland ice caps between 1988 and 1993, contain particularly valuable climate data. They demonstrate the same late-glacial climatic fluctuation as biotic databases, confirming that warming was particularly robust. According to Alley et al. (1993), the most recent of these-which marked the shift from the Younger Dryas stadial to the Holocene interglacial proper-implicated a 7°C rise in temperature in less than 50 years. The sporadic on and off swapping of the thermohaline-conveyor belt-movement in the Atlantic Ocean may be correlated to these abrupt climate changes. For instance, a selective close-down of the Atlantic conveyor due to large oscillations of meltwater from nearby Northern pole glaciers most probably triggered the Younger Dryas climatic reversal.

7.4. THE CLIMATIC OPTIMUM OF THE EARLIER HOLOCENE

Several Surface of the earth processes had lagged responses because climate changes were particularly intense during the thousands of years either part of the Pleistocene-Holocene border. Edwards et al. in 1993 explained about, for instance, worldwide "eustatic" sea levels were around 40 meters at 11,000 cal. yBP and only achieved modern levels about 6,000 to 5,000 years earlier. Likewise, many landscapes witnessed a time of momentary geomorphic destabilization during the Pleistocene-Holocene transformation, with high levels of erosion and sediment flux, even though soil configuration and vegetation canopy frequently declined behind the climatic changes.

From the enlargement of species like-the pond tortoise and the water chestnut-north of their current European climatic constraints, a thermal optimal level can be detected only about 10,000.0 and 6,000.0 cal. yBP in many mid-latitude Northern polar zones.

Changes in the Planet's receiving of solar radiation, which are astronomically operated, are the main reason for these long-term climatic conditions as illustrated in Figure 7.2. One of the three Croll-Milankovitch cycles "the precession of the equinoctial point" was at the opposite end of the cycle at the beginning of the Holocene than it is now. As a result, the North Pole received at least 8% more solar energy during the summer season than it does now, and northern summer days were hotter.

Relict landscapes can provide information about past climates since climatic patterns have such a significant impact on geomorphic structures. Lakes and Desert dunes, which are found in the sub-tropics and tropics, are two of the most climate-sensitive geomorphic regimes. Fossil mounds have assisted to have an insight into the altering extent and position of the globe's arid zones because of the intimate ties between eolian procedures and climatic aridity-Goudie (1983). Eolian landscapes unveil not only seasonal climate aridity, and also past wind directions and strengths. High-level lakes, on the other hand, reflect durations of a positive balance of water, which are correlated with time frames of wetter climate in the Holocene. Shoreline terraces and other associated landforms that went over current lakes provide direct geomorphic proof of erstwhile high levels.

Dunes were not always as arid as they are today, it was recognized in the early 20th century. Aside from ancient Saharan rock paintings portray flocks of savanna vertebrates and proof of high-level lakes, whilst rodent roosts and pollen figures denote much better, frequently wooded vegetation shield. The last "pluvial" time frame in the tropical regions began during the concluding Pleistocene and lasted throughout the first half of the Holocene-14,000 to 5,500 (cal. yBP)-according to ^{14}C dating. Street-Perrott et al. in 1991 discussed about the rainy season in northwest India, the Sahara, and East Africa increased by between 150 to 400 millimeters annually, according to water-balance estimations for these magnified initial Holocene lakes. A boosted monsoon flux ushered summer rains much farther north than they are now, resulting in this rise.

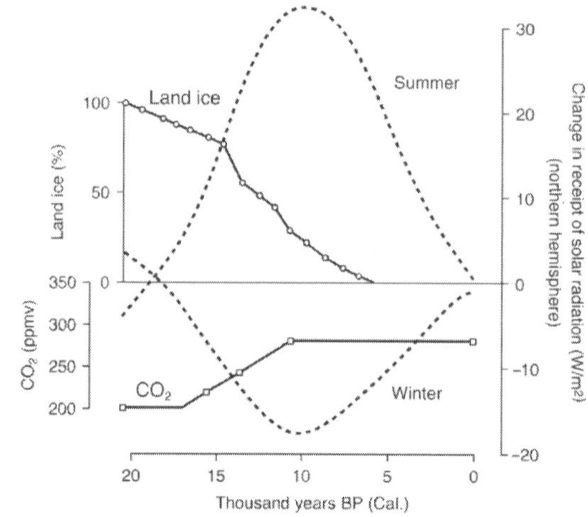

Figure 7.2. From the recent glacial maximum, the Northern Hemisphere's solar radiation and other climatic factors have changed. (Kutzbach and Street-Perrott, 1985).

Source: https://www.nature.com/articles/317130a0.

The greater monsoon was triggered by the-thermal optimum-which resulted in increased heating above the land-particularly in south-Asia-and low pressure, that trapped in humid and rainy seasonal air currents from the southern pole.

At this time, a changed sequence of climate change emerged in mid-latitude territories like the North American plain. The midwestern plain forest border shifted east-wards till 8,000 cal. yBP, according to pollen statistics, but then moved back the other way. Because moisture accessibility controls this limit now, it is sensible to be used as a proxy climatic index, recording warmer climate patterns amid 8,000 to 4,500 cal. yBP. Such climatic segment is also recorded in sedimentary rocks from nearly of the several small streams of the North American Rocky Mountains. According to Haskell et al. in 1996 described, for instance, geo-chemical, and di-atom assessments reveal high salt content throughout the first 50% of the Holocene, whereas reservoir water was pretty fresh beforehand and afterward this duration.

The ecosystem seems to have intensified the orbital pushing of the early Holocene climate-related optimum. According to modeling experimentation, by modifying the surface moisture circumstances and albedo, vegetation had a major impact on climate patterns in crucial boundary areas, like the rim of forest and dunes biomes. Foley et al. (1994); and Gasse (2000) estimates showed that the broadened forest influenced an additional 4°C warming in spring temperature all along boreal tundra ecotone 6,000 to 7,000 years earlier, in regards to that mainly caused by solar radiations.

7.5. CLIMATE VARIABILITY IN THE LATE HOLOCENE

Despite being less pronounced than in the before time Holocene, the subsequent Holocene saw some substantial climatic changes. Thar deserts, the Arabian, and the modern Saharan were formed as a result of continuous cooling succeeding the Holocene thermal optimized and climatic drying up of the northern sub-tropics following 6,000 cal. yBP. Street-Perrott et al. (1991); and Gasse (2000) demonstrated that at this moment, the water levels of certain low latitude reservoirs dropped sharply (Figure 7.3), indicating that the rainy-season circulation system, which regulates tropical heavy rains, was being suppressed quickly.

Short-time nationalist alterations from warm to cold, or dry to wet, documented in the rocky layers of peat bogs or the development and pull back of mountain ice caps, were overlaid on longer period climatic patterns linked to orbital siriasis forcing (Charman et al., 1999). The issue of whether these fluctuations portray true, consistent cycles, as well as the sources of these climate changes, remains unanswered. Variations in the solar's radiation output, related to sun-spot oscillations, and the heat protecting impact of volcanic activity with huge dust outflows are two of the utmost governing procedures for sub-Milankovitch climatic changes.

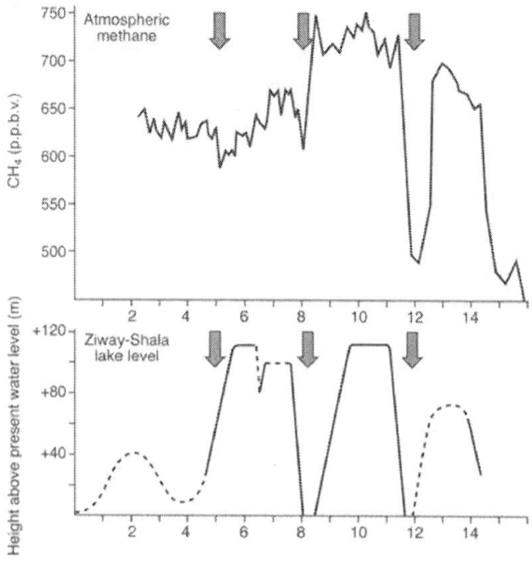

Figure 7.3. A record of tropical African lake levels is correlated to the concentrations of atmospheric methane found in a Greenland core of ice (Roberts, 1998; Gillespie et al., 1983; Blunier et al., 1995). Lengthier climate patterns are interrupted by abrupt arid occurrences, as seen by the arrows. The times are in 1000.0 yBP (cal).

Source: https://link.springer.com/referenceworkentry/10.1007%2F978-1-4020-4411-3_102#:~:text=Ocean%2Datmosphere%20interactions%20control%20 Earth's,on%20Earth%20is%20solar%20radiation.

Volcanic can have a significant impact on the planet's climate, for instance, by forming a dust mask that minimizes the atmosphere's clarity. Following important historical volcanic activities, like the one at Laki in Ice-land in 1783 A.D (Grattan and Brayshay, 1995), modern writers like Benjamin Franklin noted intense late cold snaps and strange hazes in areas far off the volcanic eruptions. Zielinski et al. considered a few prior Holocene explosions were much wider than the Laki fissure outbreak, and it has been speculated that they had a substantially larger climatic influence. The Aegean is the land of Santorini "Thera," for instance, burst disastrously in around 1628 B.C.-3,578 cal. yBP-compelled the Bronze Age populace to run away or die. The tree of Irish bog oaks was exceptionally limited at this moment, according to dendro-climatological research, whereas bristlecone pine growing rings of a similar age in California display symbols of rime damage. These have been attributed to a more temperate or wet seasons climate. Nevertheless, Kuniholm et al. in Turkey, where summers are hot and dry, the same global climate change seems to have resulted in more favorable growing conditions and wider tree rings, rather than fewer. Although this research connecting volcanic eruptions, climate, and tree rings is compelling, it also implies that the climate change was typically brief, lasting only a decade or fewer in certain instances.

According to Bond et al. (2001), alterations in solar radiation are also being known to play a role in Holocene climate conditions, usually based on the analysis that cyclic oscillations in climate measures have regularities similar to those originating in the database of the solar task. Mauquoy et al. (2002) Well-dated climatic irregularities have been associated with a renowned forcing role in certain circumstances, like European temperature minimal combined with the Maunder lowest in solar pursuit from 1650–1715 A.D.

Besides exterior factors, cyclic procedures within the planet's surface, like the quasi-cyclic approximately 1,500 years Bond rounds, in which ice sheet-armadas-floated south-ward towards the North Atlantic, delaying the oceanic thermo-haline conveyer belt, will have influenced Holocene weather. The famous ENSO-El Niño-Southern Oscillation-is a similar but much greater frequency auto-cyclic occurrence. As stated by Gagan et al. even though ENSO-like happenings seem to have occurred all across the Holocene, scientific proof suggests that the contemporary ENSO frequency band was not formed till the mid-Holocene, based on high-resolution firm isotope assessment of corals.

The previous millennium of the Holocene has the greatest-resolved and utmost accurately optimized climate data, counting both the Little Ice Age (approximately AD 1250 to 1850, especially AD 1590 to 1850) and the Medieval Climate Oddity (about AD 1000 to 1250) when main rivers such as the Tiber and the Thames frequently froze. Data comes from proxy-climate databases, such as historical observations, instrumental data, and tree rings for the last 2-century. A variety of synthetic curves for Northern Hemisphere changes in temperature have been created from existing data sets, after Mann et al.'s (1999) initial work. These have a "hockey-stick" structure, with temperatures falling gradually to a low point in the 17th century before dramatically rebounding over the last 150 years. How much proportion of the reported rise of 1°C since the mid-19th century is attributable to manmade growth in greenhouse gases (GHGs), and how much of that is due to natural fluctuation, is still an open subject. Furthermore, global warming is now likely greater than they have ever been in the last 1,000 years as well as the majority of the warming since 1980 is most certainly due to human activity (Verschuren et al., 2004).

7.6. VARIATION OF HOLOCENE TREELINE

7.6.1. Treeline and Timberline

Climatically treelines are sensitive transition regions among the alpine and closed forest, arctic, or dry grassland habitats. Precipitation and Temperature thresholds, in most situations, influence whether or not a place is treeless (Arno and Hammerly, 1993; Körner, 1999). Several other elements, including flora type, snow cover, soil type, wind, and topography can co-determine the tree growth border locally. The treeline spatial position is defined by the definition of "tree," which is independent of the abiotic and biotic factors that generate this prominent vegetation barrier. A minimum height of 8 to 2 meters is the greatest widely used criterion for defining "tree" at treeline (Arno and Hammerly, 1993; Körner, 1999). Some researchers choose to differentiate between treeline and timberline in their research (e.g., utilizing the height of the tree as a criterion of diagnosis: for the treeline, it is 2 m and for timberline, it is 8 m). The transition region between treeless plant communities and closed forests is not consistently broad: drought-caused treelines and polar treelines can generate very vast transition regions like parks and gardens with broadly spread plants (Arno and Hammerly, 1993). Mountain treelines, on the other hand, have relatively small transitional regions (i.e., 100.0–200.0 m vertical length). This range may coincide with the above-mentioned altitudinal range among treeline and timberline (8 versus 2 m tree height, Tinner and Theurillat, 2003).

7.6.2. Reconstructing Variations in the Treeline

The treeline reconstruction fluctuations can be done in a variety of ways. The most essential approaches are stomata, pollen, megafossils (particularly tree rings), and macrofossils. The highest treeline positions can also be determined using soil, charcoal, and phytolith analyses. The spatial resolutions of the approaches differ. Pollen data have poor spatial resolutions because pollen is easily removed, carried, and deposited through winds across great distances. As a result, relying solely on pollen to recreate treeline changes is problematic. For reconstruction of timberland, megafossil, and macrofossil analyses are thought to be more credible (Birks, 2001; Tinner and Theurillat, 2003). because they offer better taxonomic and geographical resolutions than pollen studies (meters to kilometers). Megafossils and m Macrofossils have the drawback of not being as common in Holocene deposits as pollen. Treeline reconstructions need records dating back to the early glacial period from contemporary treelines. There are two prerequisites: (a) the treeline data' chronology should be established to an absolute level with enough 14°C-dates of land fossils, (b) the study areas should be close to the current treeline location.

Holocene treeline reconstructions variations must be compared to present treeline locations to determine the degree of previous climate changes. It is considered that modern tree distribution is in equilibrium with the environment of the previous few decades when estimating past temperature variations. Individual trees in the remote region affected by substantial human impact (i.e., the Alps) may nevertheless signify the probable altitudinal limitation of tree and forest development. It is probable to transform previous treeline variations (m) into the (summer) changes in temperature (C) utilizing contemporary lapse rates (according to the supposition of stabilizing Holocene) (Haas et al., 1998), even though greater treelines in the previous era manifested relatively warm durations that were reliable so to enable subalpine trees to relocate up-slope (100.0–200.0 years of response lags). We concentrate on just a small well yet typical case study from controlled temperature hill treelines in Northern America and Europe, in an approach to synthesizing similar trends over two continents.

7.6.3. Dynamics of Holocene Treeline

Since Holocene treeline variations reconstructions represent previous local dynamics, direct comparisons between sites are required to determine larger-scale patterns which may be linked to climate change. macrofossil and Pollen stratigraphies inside the Italian and Swiss Alps enabled the identification of cool (wet) and warm (dry) phases (Wick and Tinner, 1997), which are consistent with alpine foreland lake-level variations (CE-1 to CE-8, Haas et al., 1998) as well as oscillations of the glacier (Figure 7.4). Macrofossils and Pollen from mire deposits and lakes across altitudinal transects allow treeline changes across the Holocene to be reconstructed (Tinner and Theurillat, 2003). The locations of treeline and timberline fluctuated inside a band of 100.0–180.0 m through the Holocene, and the treeline was at its highest point (ca. 180.0 m greater than now in the era 10,000–6,000 cal. yBP, according to these reconstructions (Figure 7.5). In cold (humid) seasons, massive diebacks of treeline vegetation happened (CE-1 to CE-8 and within the small Ice Age; Figures 7.4). From 8200.0 to 7500.0 cal. yBP (CE-3), a dramatic treeline drop occurred, with two minima lying at 8,000 and 7,600 cal. yr B.P. (Figure 7.4, sites Gouillé Rion and Lengi Egga). Treeline and Timberline dropped by about 180.0 m and 300.0–400.0 m, correspondingly, around 6000.0–5000.0 cal. yBP (Figure 7.5).

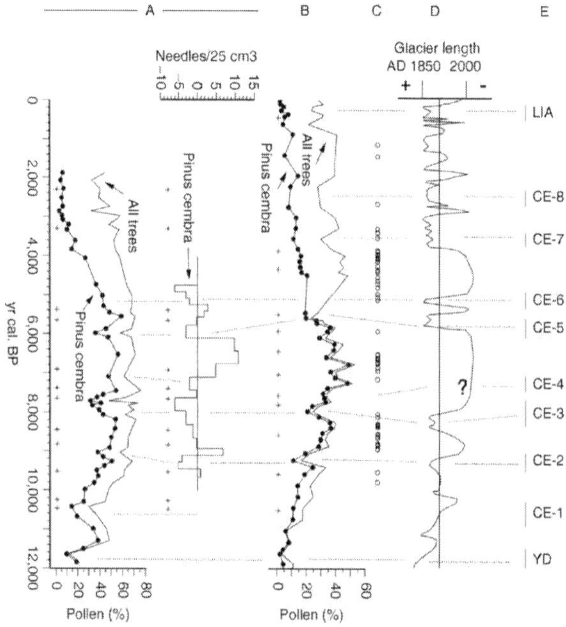

Figure 7.4. Correlation among macrofossil and pollen reports of great elevation glacier and places variations through the former 12,000 years; (A) percentages of pollen and Gouille′ Rion macrofossil concentrations (Swiss Alps); (B) percentages of pollen at Lengi Egga (Swiss Alps); (C) radiocarbon wood dates and debris of organic accumulated in Alpine glaciers front (Hormes et al., 2001); (D) Swiss glaciers determined length change (Maisch et al., 1999); (E) cold-humid phases of central European (Haas et al., 1998). +in a and b determine the chronological location of radiocarbon dates worked for depth-age models.

Source: https://link.springer.com/referenceworkentry/10.1007%2F978-1-4020-4411-3_102#:~:text=Ocean%2Datmosphere%20interactions%20control%20Earth's,on%20Earth%20is%20solar%20radiation.

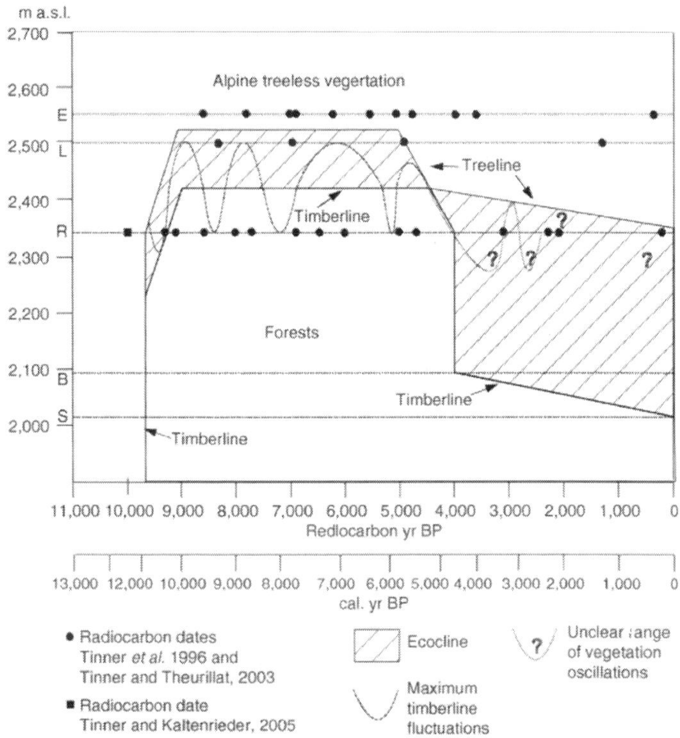

Figure 7.5. Depending on pollen sequences and radiocarbon-dated macrofossil, the estimated treeline elevation (m) and Holocene timberline in the central Alps of Swiss (Tinner and Theurillat, 2003).

Source: https://www.tandfonline.com/doi/full/10.1657/1523-0430%282003%29035%5B0158%3AULEAFO%5D2.0.CO%3B2.

The human effect was primarily responsible for the decline of highly forested regions (timberline) within the Alps over the last 5,000 years, while the path of treeline provides an even more accurate estimate of climate impact (Figure 7.4). It is likely to calculate the temperature oscillations range of Holocene in the Alps to 0.80–1.20°C around 10,500.0 to 4,000.0 cal. yBP, where mean (summer) temperatures are around 0.8–1.2°C greater than now, using a constant lapse rate of 0.70°C/100.0 m. Summertime temperatures gradually decreased by about 4,000 cal. yBP, reaching levels similar to that of the 20th century.

The Holocene tree line changes in the Alps followed a pattern comparable to that reshaped for Scandinavian places. Treeline reached its topmost limit about 9500.0 (–9000) to 6500.0 (–6000) cal. yBP within Sweden and Norway, around 200.0–400.0 m over modern treeline location, as according to pollen, macrofossil, and megafossil studies (Dahl and Nesje, 1996; Kullman, 1995; Barnett et al., 2001; Barnekow, 1999). Around 8300.0 and 8000.0 cal. yBP, treelines dropped slightly at certain study locations (Barnekow, 1999; Barnett et al., 2001; Dahl and Nesje, 1996). Temperatures were 1.30–2.00°C hotter than now at the time of peak treeline possibility, according to today's lapsed rates.

Pollen profiles give multi-millennial year data in western North America of alpine timberline changes. Such data were mostly collected from deposited sediments in highly elevated bogs and lakes, but in most situations, the records of sedimentary' limited date control limit their use to just reconstructing overall changes in treeline location during the Holocene. The analysis of tree rings obtained from dead snags or living trees close to the timberline of alpine in the Cordillera (North American) has revealed a more precise depiction of timberline changes over the last 1,000 years. The timberline of alpine remained higher than it was in the earlier to mid-Holocene as well as a widespread and general drop was begun by ca. 6000.0–5000.0 cal. yBP, according to most pollen studies (Hebda, 1995). Numerous distinct proxies for assessing historical climatic and environmental conditions have confirmed pollen-based estimations for the position of the Holocene timberline. These findings show that the average blooming season temperature throughout western North America was 1–4°C higher than it was in the slightly earlier mid-Holocene (Figure 7.6).

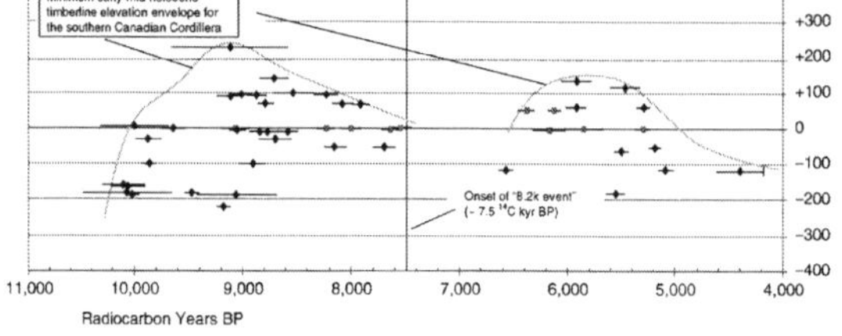

Figure 7.6. Depending on wood macrofossils of radiocarbon-dated retrieve from high-elevation places, this is the approximate minimal earliest mid-Holocene hillside height in the Canadian Cordillera (southern) (Reasoner et al., 2001).

Source: https://journals.sagepub.com/doi/abs/10.1191/0959683016802235 67.

Macrofossils of radiocarbon-dated maintained in alpine habitats are probably the best direct and clear evidence for historical great stands of alpine timberline throughout western North America.

These data are collected in some cases from plentiful conifer needles maintained within lacustrine sediments from areas presently beyond treeline that is natural (Reasoner and Hickman, 1989), but in others from logs (megafossils) maintained in soil profiles and alpine bogs (Kearney and Luckman, 1983; Carrara et al., 1991). Likewise, earlier Holocene fossil wood has been discovered at the termini of a variety of glaciers in Canada(western) (Luckman, 1988), indicating that earlier recognized forest of Holocene once occupied regions currently occupied by ice.

The discovery of wood dating from 10,000 to 5,000 years ago at places 100–200 meters just above the current timberline strongly suggests that subalpine woodlands existed at this elevation throughout the earlier Holocene. Reasoner et al. (2001) gathered early-mid Holocene macro-fossil data from all over southwestern Canada and mapped it against contemporary timberline.

The findings show that there were two distinct eras of higher-than-present timberline throughout the earlier Holocene: one from the start of the Holocene to around 8,200 cal. yBP (7500 14C yBP) and another around 7,300 and 5,700 cal. yBP (6400 and 5000 14C yBP, Figure 7.6). These findings are consistent with European data (i.e., Figure 7.5), which show that the timberline was greater than during the earlier mid-Holocene, implying that the 8.2 events influenced alpine treelines across western North America. To remedy this problem, more high-resolution data for Cordillera of North American are needed.

Dendrochronological investigations of surviving trees close to the alpine timberline and in situ deceased snags have revealed numerous small-scale changes in the alpine timberline over the previous millennium. For instance, the sudden synchronous rescission of 92 tree-ring documents linked to the 1690s "harsh winter" (Luckman, 1994), distinctly shows that high changes to the alpine timberline also happened on decadal timelines and that these variations have indeed been influenced by natural variation of the territorial climate (Figure 7.7).

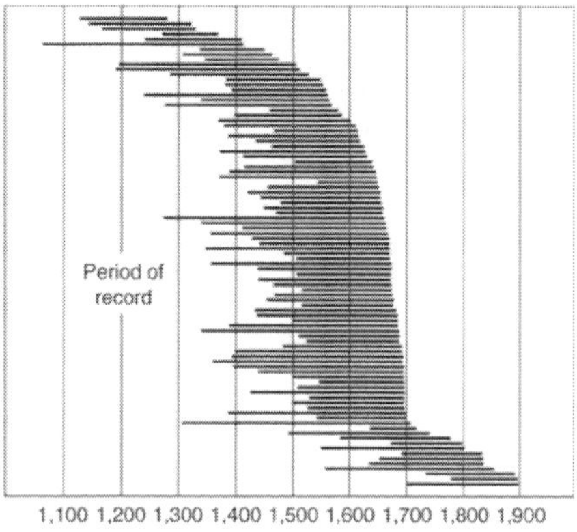

Figure 7.7. Calendar-dated snags discovered from the Site of Athabasca Glacier maintained the duration of the record. Fallen and Standing snags can be found near and over the current treeline, or if the snags are much larger than that of the trees that are currently growing in the area. The Athabasca Glacier's significant Small Ice Age advancements have been timed to 1714 (Luckman, 1994).

Source: https://link.springer.com/article/10.1007/BF01092412.

7.6.4. Creating a Link Between the Past and the Future

The composition and position of the ecotone timberline have been influenced by Holocene climatic change. The trends of millennial-scale show a progressive drop in insolation of the Northern Hemisphere, which was around 11% greater than it is now in the earlier Holocene. Higher-frequency oscillations attributable to variations in volcanic activity, solar irradiance, and ocean circulation or a combined effect of these variables are superimposed on the long trend. Manmade climate change will probably exceed the hottest Holocene circumstances during the next 100 years. Issues regarding global warming include changes in alpine and boreal life, as well as extinctions, collapses, and migrations. Interestingly, limited studies have looked at how ecosystems like treeline communities have responded to climatic variation in the past. One cause for this is that effective investigations need independent climate proxies and also extremely great temporal resolution (10.0–20.0 years/sample). High-resolution investigations, especially macrofossil research, are critically needed to understand how treelines might react to changing climate.

REFERENCES

1. Alley, R. B., Meese, D. A., Shuman, C. A., Gow, A. J., Taylor, K. C., Grootes, P. M., White, J. W.C., Ram, M., Waddington, E. D., Mayewski, P. A., & Zielinski, G. A., (1993). Abrupt increase in Greenland snow accumulation at the end of the Younger Dryas event. *Nature, 362*, 527–529.
2. Arno, S. F., & Hammerly, R. P., (1993). *Timberline-Mountain and Arctic Forest Frontiers* (p. 304). Seattle: The Mountaineers.
3. Bard, E., & Broecker, W. S., (1992). *The Last Deglaciation: Absolute and Radiocarbon Chronologies* (Vol. 1, pp. 1–23). NATO ASI Series. Berlin: Springer.
4. Barnekow, L., (1999). Holocene tree-line dynamics and inferred climatic changes in the Abisko area, northern Sweden, based on macrofossil and pollen records. *Holocene, 9*, 253–265.
5. Barnett, C., Dumayne-Peaty, L., & Matthews, J. A., (2001). Holocene climatic change and tree-line response in Leirdalen, central Jotunheimen, south central Norway. *Rev. Paleobot. Palynol., 117*, 119–137.
6. Birks, H. H., (2001). Plant macrofossils. In: Smol, J. P., Birks, H. J.B., & Last, W. M., (eds.), *Tracking Environmental Change Using Lake Sediments* (pp. 49–74). Dordrecht: Kluwer.
7. Blunier, T., Chappellaz, J., Schwander, J., Stauffer, B., & Raynaud, D., (1995). Variations in atmospheric methane concentration during the Holocene epoch. *Nature, 374*, 46–49.
8. Bond, G., Kromer, B., Beer, J., Muscheler, R., Evans, M. N., Showers, W., Hoffmann, S., Lotti-Bond, R., Hajdas, I., & Bonani, G., (2001). Persistent solar influence on north Atlantic climate during the Holocene. *Science, 294*, 2130–2136.
9. Bradley, R. S. and Jones, P. D., (1992). *Climate Since A.D. 1500* (Vol. 1, pp. 1–20). London: Routledge.
10. Carrara, P. E., Trimble, D. A., & Rubin, M., (1991). Holocene treeline fluctuations in the Northern San-Juan Mountains, Colorado, USA, as indicated by radiocarbon-dated conifer wood. *Arct. Alp. Res., 23*, 233–246.
11. Charman, D. J., Hendon, D., & Packman, S., (1999). Multi-proxy surface wetness records from replicate cores on an ombrotrophic mire: Implications for Holocene Paleoclimate records. *J. Quaternary Sci., 14*, 451–463.

12. COHMAP Members, (1988). Climatic changes of the last 18,000 years: Observations and model simulations. *Science, 241*, 1043–1052.
13. Dahl, S. O., & Nesje, A., (1996). A new approach to calculating Holocene winter precipitation by combining glacier equilibrium-line altitudes and pine-tree limits: A case study from Hardangerjokulen, central southern Norway. Holocene, 6, 381–398.
14. Edwards, R. L., Beck, J. W., Burr, G. S., Donahue, D. J., Chappell, J. M.A., Bloom, A. L., Druffel, E. R.M., & Taylor, F. W., (1993). A large drop in atmospheric $^{14}C/^{12}C$ and reduced melting in the Younger Dryas, documented with ^{238}Th ages of corals. *Science, 260*, 962–968.
15. Foley, J. A., Kutzbach, J. E., Coe, M. T., & Levis, S., (1994). Feedbacks between climate and boreal forests during the Holocene epoch. *Nature, 371*, 52–54.
16. Gagan, M. K., Ayliffe, L. K., Beck, J. W., Cole, J. E., Druffel, E. R.M., Dunbar, R. B., & Schrag, D. P., (2000). New views of tropical paleoclimates from corals. *Quaternary Sci. Rev., 19*, 45–64.
17. Gasse, F., (2000). Hydrological change in the African tropics since the last glacial maximum. *Quaternary Sci. Rev., 19*, 189–212.
18. Gillespie, R., Street-Perrot, F. A., & Switsur, R., (1983). Post-glacial arid episodes in Ethiopia have implications for climate prediction. *Nature, 306*, 681–683.
19. Goudie, A. S., (1983). The arid Earth. In: Gardner, R., & Scoging, H., (eds.), *Megageomorphology* (pp. 152–171). Oxford: Clarendon Press.
20. Grattan, J., & Brayshay, M., (1995). An amazing and portentious summer: Environmental and social responses in Britain to the 1783 eruption of an Iceland volcano. *Geogr. J., 161*, 125–134.
21. Gulliksen, S., Birks, H. H., Possnert, G., & Mangerud, J., (1998). The calendar age of the Younger Dryas – Holocene transition at Kråkenes, western Norway. *The Holocene, 8*, 249–260.
22. Haas, J. N., Richoz, I., Tinner, W., & Wick, L., (1998). Synchronous Holocene climatic oscillations recorded on the Swiss Plateau and at timberline in the Alps. *Holocene, 8,* 301–309.
23. Haskell, B. J., Engstrom, D. R., & Fritz, S. C., (1996). Late quaternary paleohydrology in the North American Great Plains inferred from the geochemistry of endogenic carbonate and fossil ostracods from Devils Lake, North Dakota, USA. *Paleogeogr, Paleoclimatol, Paleoecol., 124*, 179–193.

24. Hebda, R. J., (1995). British Columbia vegetation and climate history with a focus on 6 ka BP. *Géographie Physique et Quaternaire, 49*, 55–79.
25. Hormes, A., Müller, B. U., & Schlüchter, C., (2001). The Alps with little ice: Evidence for eight Holocene phases of reduced glacier extent in the Central Swiss Alps. *Holocene, 11*, 255–265.
26. Kearney, M. S., & Luckman, B. H., (1983). Holocene timberline fluctuations in Jasper National Park, Alberta. *Science, 221*, 261–263.
27. Körner, C., (1999). *Alpine Plant Life* (p. 338). Berlin: Springer.
28. Kullman, L., (1995). Holocene tree-limit and climate history from the Scandes-mountains, Sweden. *Ecology, 76*, 2490–2502.
29. Kuniholm, P. I., Kromer, B., Manning, S. W., Newton, M., Latini, C. E., & Bruce, M. J., (1996). Anatolian tree rings and the absolute chronology of the eastern Mediterranean, 2220–718 BC. *Nature, 381*, 780–783.
30. Kutzbach, J., & Street-Perrott, F. A., (1985). Milankovitch forcing of fluctuations in the level of tropical lakes from 18 to 0 kyr BP. *Nature, 317*, 130–134.
31. Luckman, B. H., (1988). 8000-year-old wood from the Athabasca Glacier, Alberta. *Can. J. Earth Sci., 25*, 148–151.
32. Luckman, B. H., (1994). Evidence for climatic conditions between ca. 900–1300 A.D. in the Southern Canadian Rockies. *Clim. Change, 26*, 171–182.
33. Maisch, M., Wipf, A., Denneler, B., Battaglia, J., & Benz, C., (1999). The glaciers of the Swiss Alps. Glacier high level 1850, current glaciation, glacier retreat scenarios. Zurich: vdf, p. 373.
34. Mann, M. E., Bradley, R. S., & Hughes, M. K., (1999). Northern Hemisphere temperatures during the past millennium: Inferences, uncertainties, and limitations. *Geophys. Res. Lett., 26*, 759–762.
35. Mauquoy, D., van Geel, B., Blaauw, M., & Plicht, van der J., (2002). Evidence from northwest European bogs shows "little ice age" climatic changes driven by variations in solar activity. *The Holocene, 12*, 1–6.
36. Mörner, N.-A., (1976). The Pleistocene/Holocene boundary: A proposed boundary-stratotype in Gothenberg, Sweden. *Boreas, 5*, 193–275.

37. Reasoner, M. A., & Hickman, M., (1989). Late quaternary environmental change in the Lake Ohara Region, Yoho-National-Park, British Columbia. *Paleogeogr. Paleoclimatol. Paleoecol., 72*, 291–316.
38. Reasoner, M. A., Davis, P. T., & Osborn, G., (2001). Evaluation of proposed early-Holocene advances of alpine glaciers in the North Cascade Range, Washington State, USA: Constraints provided by paleoenvironmental reconstructions. *Holocene, 11*, 607–611.
39. Roberts, N., (1998). *The Holocene. An Environmental History*. Oxford/Boston: Blackwell.
40. Ruddiman, W. F., Street-Perrott, F. A., & Bartlein, P. J. (1993). *Global Climates since the Last Glacial Maximum* (pp. 415–467). Minneapolis: University of Minnesota Press.
41. Street-Perrott, F. A., Mitchell, J. F.B., Marchand, D. S., & Brunner, J. S., (1991). Milankovitch and albedo forcing of the tropical monsoons: A comparison of geological evidence and numerical simulations for 9,000 yr BP. *Trans. R. Soc. Edinb., Earth Sci., 81*, 407–427.
42. Stuiver, M., & Reimer, P. J., (1993). Extended ^{14}C database and revised CALIB radiocarbon calibration program. *Radiocarbon, 35*, 215–230.
43. Tinner, W., & Kaltenrieder, P., (2005). Rapid responses of high-mountain vegetation to early Holcene environmental changes in the Swiss Alps. *Journal of Ecology, 93*(5), 936–947.
44. Tinner, W., & Theurillat, J.-P., (2003). Uppermost limit, extent, and fluctuations of the timberline and treeline ecocline in the Swiss Central Alps during the past 11,500 years. *Arct., Antarct., Alp. Res., 35*, 158–169.
45. Verschuren, D., Briffa, K., Hoelzmann, P., Barker, P., Scott, L., Barber. K., Snowball, I., & Roberts, N., (2004). Holocene climate variability in Europe and Africa: A PAGES PEP III time-stream 1 synthesis. In: Battarbee R. W., Gasse F., & Stickley C. E., (eds.), *Past Climate Variability Through Europe and Africa* (pp. 567–582). Dordrecht: Kluwer.
46. Watson, R. A., & Wright, H. E. Jr., (1980). The end of the Pleistocene: A general critique of chronostratigraphic classification. *Boreas, 9*, 153–163.
47. Webb III, T., Bartlein, P. J., Harrison, S. P., & Anderson, K. H. (1993). Vegetation, lake levels, and climate in eastern North America for the past 18,000 years. *Global climates since the last glacial maximum*, 415-467.

48. Wick, L., & Tinner, W., (1997). Vegetation changes and timberline fluctuations in the Central Alps as indicator of Holocene climatic oscillations. *Arct. Alp. Res., 29*, 445–458.

49. Zielinski, G. A., Mayewski, P. A., Meeker, L. D., Whitlow, S., Twickler, M. S., Morrison, M., Meese, D. A., Gow, A. J., & Alley, R. B., (1994). Record of volcanism since 7000 B.C. from the GISP2 Greenland ice core and implications for the volcano-climate systems. *Science, 262*, 948–952.

Chapter 8

Human Evolution and Climate Change

CONTENTS

8.1. Introduction 204
8.2. The Family Tree of Hominins 205
8.3. Hominin Attributes' Evolution 207
8.4. Worldwide Climate Change 209
8.5. Climate and Evolution 210
8.6. What Does Paleoclimatology Teach Us? 212
8.7. Conclusions and Summary 213
References 214

8.1. INTRODUCTION

The essential questions that have always bloomed interest and seized our attention are why, how, and where humans evolved. Humans journeyed to various regions of the world beginning as soon as 2.0 Ma after evolving in Africa that was verified through developments in DNA study and several new hominin fossil findings as ancient as 6 Ma. The family Hominidae consists of fossil or living members called Hominins. Questions have been raised after finding proof of their ancient migration from Africa. What was the reason that the security of trees was left by humankind? Bipedalism gave rise to the advancement of species that were unsuccessful while only one remained, so what was the motivation that pushed the hominins towards it? Did the paleoenvironmental pressures receive direct or passive comebacks in the form of brain advancement that generated imagination, language, and art, the evolvement of tools, and the migration from Africa?

A lot of information has been produced on paleoclimate and climate over the previous few decades as the necessity to assess the possible danger of global warming and understanding our planet's climate system has risen. The examination of future, present, and past climate is made possible by facilitating computer models called global climate models or GCMs that were borne through research. Though the details of global climate transformation, the versatile approach combining biologic, anthropologic, and geologic statistics is needed to comprehend the paleoclimatic and paleoenvironmental situation of human evolution.

The stone tools documentation goes back to 2.50 million years and the hominin fossil to 6.0 million years. The surfacing of the facts of paleoclimate on land is just now commencing while through the oxygen isotope documentation of abysmal sea cores, well recording has been done of the typical context of global climate at the time of the Plio-Pleistocene. Northern Tanzania, northern Kenya Turkana, Olduvai Gorge, southern Kenya, and Ologesailie are examples of some efforts to place main sites inside a wider spatial and temporal context and to manufacture inside regions (Behrensmeyer et al., 2002; Feibel et al., 1991; Brown and Feibel, 1991; Ashley and Hay, 2002; Hay, 1976). Typically, local site-based accounts complementing every find, the emphasis of most geological researches on hominin-bearing sediments has been on chronology and correlation of sites. Found from paleo-anthropological and archaeological documents, the progressive history and the geologic document present a chance to combine what we know about paleoclimate and paleoenvironment due to the rising comprehension of the significance of volcanism, astronomic, and tectonism climate forcing through the Plio-Pleistocene. There has been victory in relating human evolution to climate by a lot of scientists and to attend to the problem, a sequence of conferences and global workshops have taken place. (Potts, 1996b; Brain, 1981; Vrba et al., 1995; Behrensmeyer, 1982; Foley, 1994; Bromage and Schrenk, 1999; deMenocal, 1995; Feibel, 1997; Boaz and Burckle, 1983).

The subsequent steps will be taken to observe the relationship amid social evolution and climate:

- The hominins' phylogeny (family evolutionary past) will be studied;
- The characteristically "human" evolved features of hominins will be described;
- At the time of hominin evolution, the information about worldwide climate will be shortly summarized;
- Probable links involving social Darwinism and climate change will be examined.

8.2. THE FAMILY TREE OF HOMININS

Hominin family tree's chief design is apparent, despite the insufficient fossil documentation of hominins before the entrance of contemporary humans (possibly under a few 1,000 persons overall) (Figure 8.1). Every hominin fossil is accorded its genus term as the links among fossils is very undefined and the documentation so meager before 4 million years ago: 5.8–5.55 million yrs from Ethiopia; *Ardipithecus kadabba*, 6.0–7.0 million yrs from Chad; *Sahelanthropus tchadensis*, 4.40 million yrs from Ethiopia; Ardipithecus ramidus, and, 6.0 million yrs from Kenya; *Orrorin tugenensis*. Maybe throughout the late Miocene, in the middle of 7.0 and 9.0 Ma, the separation involving human beings and apes occurred because hominins lived at a minimum of 6 Ma.

It is possible that numerous species survived coincidentally; *A. afarensis* and *A. africanus* were significant species, and the supreme hominin family from around 4.0–2.50 Ma was the gracile australopithecines. The group *Australopithecus anamensis*, among the original australopithecines was discovered in northern Kenya, existing in a huge area extending from South

Africa to Ethiopia, dated at 4.20 Ma (Leakey et al., 1995). The contemporary human beings, brought about by the initial affiliates of the genus Homo, and robust australopithecines (for example *Paranthropus boisei*), who ultimately ceased to exist was led to the separation of 2 hominin families from the primary more gracile australopithecines at around 2.50 Ma (Figure 8.1). The lifetime of the initial homo is still undefined. 3 species were produced by the lineage's radiation: *H. rudolfensis, H. habilis,* and *H. erectus*, which means the genus was certainly existing by 1.90 Ma even though it was as ancient as 2.40 Ma (Wood, 1992).

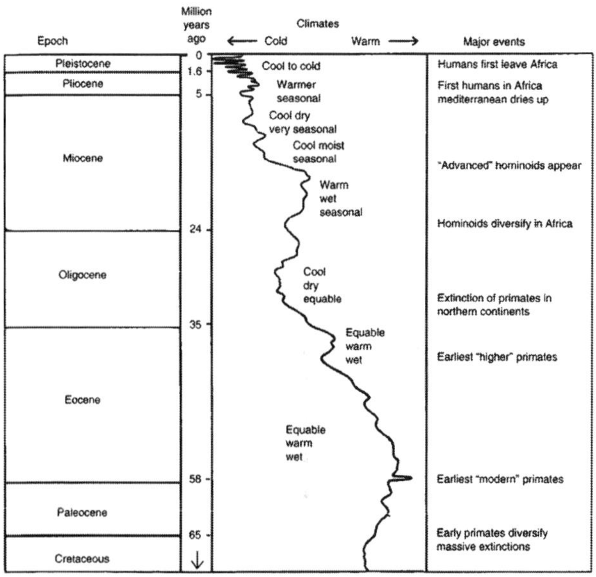

Figure 8.1. After the Pleistocene era's ice age, the course of climate, showcasing a typical culminating and cooling at the time of the Cenozoic age is outlined by the graph. Discovered in marine sediment documents, the oxygen isotope document of foraminifers helps interpret the comparative temperatures. On the right are the primate evolution's chief incidents. (Partridge et al., 1995; Tattersall, 1993).

Source: https://onlinelibrary.wiley.com/doi/abs/10.1002/ajpa.1330920312.

The genus Homo ultimately turned out to be the sole survivor after overlapping with the vigorous australopithecines in time and space. The recognition of *Homo sapiens*, i.e., functionally contemporary humans' existence has been from around 200,000 years ago (Figure 8.2).

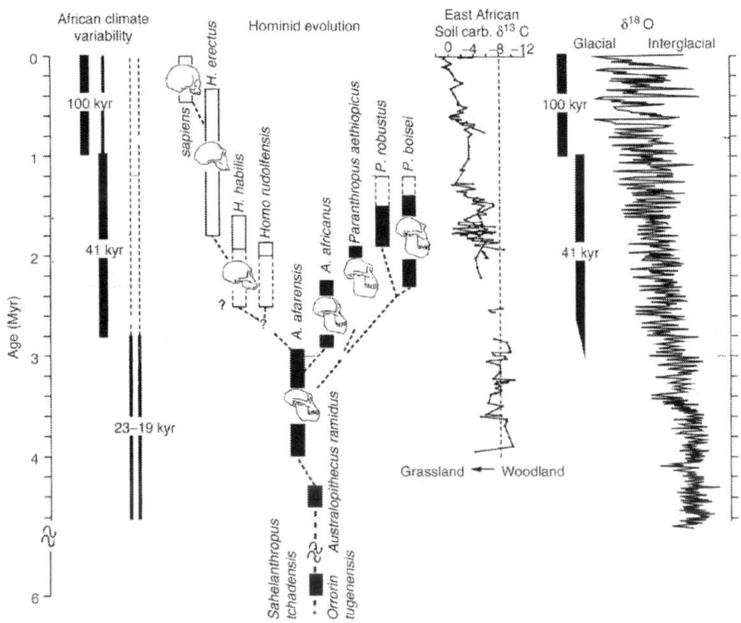

Figure 8.2. Represented through the carbon isotopes in soils' interpretation of vegetation document, the Milankovitch cycles' supremacy (far right), oxygen isotopes from the marine document's interpretation of worldwide temperatures; human being phylogeny's graph of the assessment of the African climate inconsistency (Wood, 1992; Cerling, 1992; Cerling and Hay, 1988; deMenocal and Bloemendal, 1995; Shackleton et al., 1990).

Source: https://www.nature.com/articles/355783a0.

8.3. HOMININ ATTRIBUTES' EVOLUTION

Two vital physical qualities, i.e., a) an upsurge in brain aptitude comparative to body size (related to the genus Homo) and b) bipedal locomotion (it symbolizes every hominin and came about very early) is possible to be traced in the hominin fossil document. 650.0 cm^3 in initial Homo, 400.0 cm^3 in australopithecines, and 1,400.0 cm^3 in Homo sapiens (contemporary humans) are the regular cranial aptitude limits. There seemed to be a connection in between the 2 features. The growth in brain size will take place after hominins left the security of trees and developed bipedalism as envisioned by the representations of humanoid evolution (Figure 8.3).

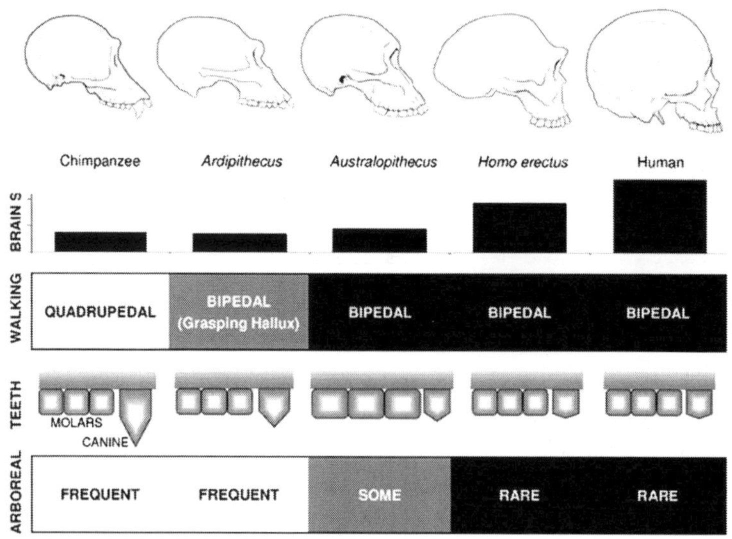

Figure 8.3. Hominin evolution's outline.

Source: https://www.nature.com/scitable/knowledge/library/overview-of-hominin-evolution-89010983/.

The brain size growth and bipedalism are self-regulating as suggested by fossil proof. *Orrorin tugenensis* (6.0 Ma), 1 of the initial hominins whose fossil proof suggests that it had already adjusted to bipedalism while it was still on the ground even though it still climbed trees (Senut et al., 2001). An "enhanced" cranial room (between 435.0 to 530.0 cm^3) but a "primitive" physique was found in a bipedal individual dated amid 2.0 and 3.0 Ma while studying *A. africanus* fossils belonging in South Africa (McHenry and Berger, 1998). A petite pelvic girdle, lengthy higher limbs, petite lower members, and a huge chest made up the primitive figure. Thus, having a bigger brain and the capability to walk on 2 legs did not mean that climbing on trees was still not significant for them.

The time the brain achieves substantial cranial volume, things like abstract thought, culture, technology, and language begin to develop. An obvious association is not suitable as sex and body size influence cranial volume. Dating back to 2.55 Ma is the primary proof of technology, i.e., the making of primitive stone tools (Semaw et al., 1997). Explicit evidence for this does not exist but Homo is presumed to be the makers of these tools. Pigments, grindstones, and stone blades which were found in old archaeological spots suggest that intricate cognitive thought and language had grown in hominins 200,000.0–300,000.0 years before though there is no record for this (McBrearty and Brooks, 2000). There are traces of various forms of art and beads found as back as 50,000.0 years ago and fish hooks, mining, and long-distance communication go back to almost 100,000 years (McBrearty and Brooks, 2000).

8.4. WORLDWIDE CLIMATE CHANGE

Since the Earth's birth over 460,000,000,0 years ago, its climate has changed drastically. Things like the Earth's orbit going through minor variations, i.e., astronomical forcing, the tectonics and the intensity of the Sun fluctuating is affecting the climate structure which is made up of 5 elements that have intricate exchanges with each other which are; atmosphere, land surface, ice, vegetation, and oceans (Ruddiman, 2001). Occurring in marine deposits ever since the evolution of hominoids almost 65,000,000 years ago, is the most whole and nonstop documentation of climate change. The analysis is going on of sets of minerals and microfossils lying in extensive sediment cores in the ocean's basins. Intended for climate change with time, some proxies such as dust blowing in from continental deserts and small marine creatures, i.e., diatoms, and foraminifers containing oxygen isotopes are employed. Moisture variation is signified by the dust meanwhile, the worldwide temperature is reflected by the isotope inconsistency (deMenocal, 1995; Hays et al., 1976). Extra material regarding climate is provided by documents of continental fragments found in soils or lakes. For instance, for determining the proportions of savannah, forests, and open woodland, present in soil carbonates is the carbon isotope which aids as a proxy for vegetation variations (Cerling et al., 1993; Cerling, 1984).

The Pleistocene era's ice ages occurred due to the drying and cooling recorded in the last 650,000,00 years, i.e., the Cenozoic period according to the terrestrial and marine documents (Figure 8.1). There were volume upsurges in worldwide ice because of moments of sudden cooling (14.0 and 2.80 Ma). 2.80 Ma signifies the beginning of Northern Hemisphere continental-scale freezing otherwise known as glaciation, while the Antarctic glaciers became everlasting and enlarged in size in 14.0 Ma. The oceanic flow was altered by the shutdown of the Isthmus of Panama which along with enormous uplifts (e.g., Western North America and Tibetan Plateau's Mountain construction) occurred due to plate tectonics which plays a part in the overall cooling of our planet (Ruddiman, 2001; Stanley, 1995).

In the current period, there is an upsurge in the scale of climate fluctuations while previously there had been swings on a variation of time scales but not as much, even though a drier and cooler climate is a long-term drift (Figure 8.1). Short-term fluctuations (solar radiance) concerning the Sun, the comparative place of the Earth is causing long-term fluctuations, i.e., the Milankovitch cycles. The planet's orbit's eccentricity produces 100 ka cycles, linked to the equinox's precession are 21 ka cycles and the incline of the planet's reflection (obliquity) by 41 ka cycles, are all part of the Milankovitch cycles. The 41 and 100 kyr Milankovitch cycles are linked to the Pleistocene ice ages which the Northern Hemisphere was dominated by (Ruddiman, 2001).

Great variability signifies the climate of the time when contemporary humans evolved from hominins about 6,000,000 years ago. The Indian monsoon structure's episodic improvement and seasonal continental circumstances including a drier and cooler climate 6,000,000 years ago (Ruddiman, 2001). The quantity of solar radiation that reaches the planet is changed by the 21-ka precession cycle which is linked to the monsoon might. The tropics where the evolution of human beings occurred were subjugated by wet-dry cycles and the lake tiers fell and rose to the pace of this most minute Milankovitch cycle. The humanoid evolutionary reaction must have been greatly influenced by this climate fluctuation which probably had a momentous impact on the primitive man.

8.5. CLIMATE AND EVOLUTION

Most of the evolution occurred in Africa which has a climatically unstable atmosphere and the fact that man became modern who can travel in space from being a dweller of trees tempts researchers to make connections amid the spectacular evolutionary variations of hominoids. Ecosystems and species would continue to be unaffected and in balance in the absence of environmental variation acting as a catalyst which is why the creation of new species is credited to climate change (Poirier and McKee, 1999; Vrba et al., 1995). It is difficult to relate a precise evolutionary modification to a precise climate condition. Significant to human existence, it is questioned if at the environmental stage, the worldwide climate features are sensed or not (Behresmeyer, 1982). A few generalizations are possible to be made even if the paleoclimate documentation is insufficient, chiefly on land and the traces of hominid fossils are extensively spread and inadequate (Figure 8.4).

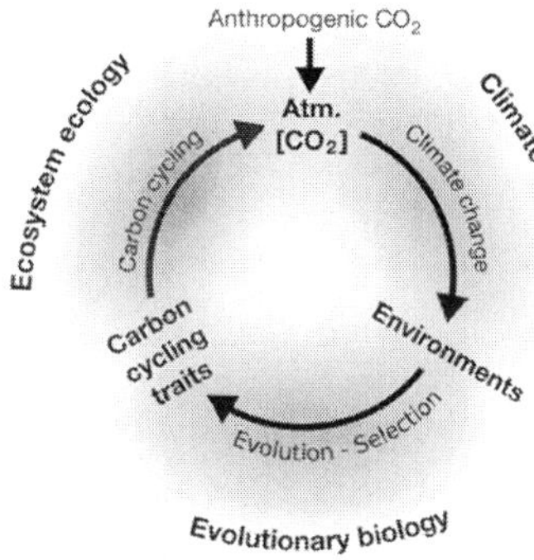

Figure 8.4. Evolution and climate change.

Source: https://phys.org/news/2018-02-climate-evolution-friends.html.

A rise in savannah plains and exposed woodlands while a decline in forests was found in Africa due to the overall drying and cooling in the last 5,000,000 –7,000,000 years. Even though hominins maintained their ability for climbing trees for a long time as indicated by fossil proof, this was the period they started to acquire bipedal locomotion. The tropics began to become dry and in the Northern Hemisphere continental glaciers began to develop about 2,500,000 years ago, in short, the Earth's climate structure went through growing changeability (Figure 8.1). Hominoid "technology" like stone devices and the genus Homo's initial entrances were close to this period coincidentally. To pursue a steadier atmosphere, their exodus from Africa is documented in about 1.70 Ma in the Republic of Georgia through the detection of Homo (hominins) and ancient tools (Gabunia et al., 2000). Africa's fluctuating environment received an evolutionary comeback from the animals according to new indications. Species tailored for grasslands were preferred by the people due to the drying and cooling climate as found in research in Ethiopia and Northern Kenya of surfing through fauna and grazing by researchers. How is evolution activated by climate variation? A filter is a role that climate adopts. Hominoid evolution's forms can be described through an ecological prototype named "variability selection" put forth by a scientist. Animals that can eagerly familiarize themselves with the latest environments are favored by the vivid changes in climate. Instead of the specialist's survival, the attention is on the generalist. Success would belong to families that can adapt to transformations and live-in diverse surroundings. On the other hand, failure to adjust to new circumstances ends in the extinction of the specialized. From tropical wildernesses to the sea ice of the arctic, sea level to great altitudes (4,500.0 m), human beings have existed in all thinkable ecological locations and endured the ice ages and famines.

8.6. WHAT DOES PALEOCLIMATOLOGY TEACH US?

The documentation of instrumental climate is not enough for us to foresee and comprehend the climate structure's variations and seasonal-to-century gauge climate inconsistency's comprehensive understanding is needed. Thus, the extensive thermometer utilization and the fluctuations in the temperature of the World earlier than 1850 and through the centuries led paleoclimatologists to construct a blueprint of it. Historic papers, additional proxy information sources, tree rings, and deposits became the basis of the assessment of the climate in the past which arose out of the primary work. Previous fluctuations in the temperature of the Earth are provided in an enhanced view now as various more paleoclimate documents are accessible from anywhere in the world today. The fluctuations in the Earth retaining the quantity of solar radiation, land, atmosphere, and ocean exchanges are related to the environmental inconsistency of the Earth which proves that climate change is natural according to the viewpoint of paleoclimates. Climate variations at a large scale are also included in the proofs present in the geologic documents. Warm surroundings at a universal level about 90–120,000,000 years ago are indicated by organisms such as corals, hot climate plants, and dinosaurs existing at great heights (Figure 8.5).

Figure 8.5. Ice cliffs facing dry-seasonal dust bands have resulted from precipitation's diverse seasonality at places such as the ice cap of Quelccaya. Extremely precise dates of archives of ice cores are provided by those layers which makes them very valuable to paleo climatologists.

Source: https://www.ncdc.noaa.gov/news/what-have-we-learned-paleoclimatology.

Around 2,000,000 years before, the climate of the past was dominated by 1,000 years old multiple ice ages, glacial retreats, and cycles of glaciers as presented by paleoclimate information. The temperature of the Earth was 9°F cooler than now about 21,000 years before when huge regions of Europe and North America were covered by gigantic sheets of ice at the time of the freshest glacial cycle's peak. There have been many variations in the climate of the planet ever since the final ice age ended over 10,000 years ago. The previous millennia are dominated by the "Little Ice Age"'s cooling and the medieval period's heating. Climate and atmosphere's current variations are brought into view by paleoclimatology. Compared to the last 80,000 years, the degree of CO_2 in the atmosphere is greater as indicated by gas bubbles imprisoned in cores of ice now. Throughout interglacial and glacial cycles, the amount of carbon dioxide was about 179 to 280.5 portions per million before human beings started to burn huge quantities of fossil fuels after the Industrial Revolution. Around 400 portions per million are the amount that CO_2 has reached today in 2014 (Study how the amount of CO_2 is traced by NOAA and regarding the Carbon Tracker). Compared to any instance in the past 1,500 years, alongside the rising of degrees of CO_2 in the atmosphere, the temperature of the globe has also increased according to numerous proxies in paleoclimate archives. In a few regions of the planet, fluctuations in the orbit of the Earth amplified its temperature over 5,000 years before and Earth has not been as hot as it is now since then. In the previous century, the temperature of the Earth has gone from being the coldest to the hottest of our present interglacial time which is a complete reversal of the cooling trend that has been there in the last 5,000 years or more which is quite extraordinary.

8.7. CONCLUSIONS AND SUMMARY

In the Hominidae family, precise evolutionary comebacks cannot be properly related to precise climatic occurrences. Nevertheless, the bipedalism acquirement and Africa's aridification (grassland and exposed woodlands' related enlargement) in the past 7.0 Ma presents a wide chance. Hominins' exodus from Africa, brain volume's growth, the stone device's (proof of technological abilities) initial arrival, and the genus Homo's primary trace is related to growing climate changeability. According to Potts (1996), animals which are favored by vivid swings in climate, can adjust to an extensive variety of environmental settings and are generalists, i.e., the "variability selection" concept may consist of the advancement of language, cognitive procedures, and art, etc., though their direct relation to climate is not possible.

REFERENCES

1. Ashley, G. M., & Hay, R. L., (2002). Sedimentation patterns in a Plio Pleistocene volcaniclastic rift-margin basin, Olduvai Gorge, Tanzania. In: Renaut, R. W., & Ashley, G. M., (eds.), *Sedimentation in Continental Rifts* (pp. 107–122). (SEPM Special Publication 73) Tulsa, OK: SEPM (Society for Sedimentary Geology).
2. Behrensmeyer, A. K., Potts, R., Deino, A., & Ditchfield, P., (2002). Olorgesailie, Kenya: A million years in the life of a rift basin. In: Renaut, R. W., & Ashley, G. M., (eds.), *Sedimentation in Continental Rifts* (pp. 97–106). (SEPM Special Publication 73) Tulsa, OK: SEPM (Society for Sedimentary Geology).
3. Behresmeyer, A. K., (1982). The geological context of human evolution. *Annu. Rev. Earth Planet. Sci., 10*, 39–60.
4. Behresmeyer, A. K., Todd, N. E., Potts, R., & McBrinn, G. E., (1997). Late Pliocene faunal turnover in the Turkana Basin, Kenya and Ethiopia. *Science, 278*, 1589–1594.
5. Boaz, N. T., & Burckle, L. H., (1983). Paleoclimatic framework for African hominid evolution. In: *SASQUA International Symposium Proceedings* (pp. 483–490). Location: Transvaal Museum, Swaziland, Africa.
6. Brain, C., (1981). Hominid evolution and climate change. *S. Afr. J. Sci., 77*, 104–105.
7. Bromage, T., & Schrenk, F., (1999). *African Biogeography, Climate Change, and Human Evolution* (p. 485). New York: Oxford University Press.
8. Brown, F. H., & Feibel, C. S., (1991). Stratigraphy, depositional environments, and paleogeography of the Koobi Fora Formation. In: Harris, J. M., (ed.), *Koobi Fora Research Project: Stratigraphy, Artiodactyls and Paleoenvironments* (Vol. 3, pp. 1–30). Oxford, UK: Clarendon Press.
9. Brunet, M., Guy, F., Pilbean, D., et al., (2002). A new hominid from the Upper Miocene of Chad, Central Africa. *Nature, 418*, 145–151.
10. Cerling, T. E., & Hay, R. L., (1988). An isotopic study of paleosol carbonates from Olduvai Gorge. *Quaternary Res., 25*, 63–78.
11. Cerling, T. E., (1984). The stable isotopic composition of modern soil carbonate and its relation to climate. *Earth Planet. Sci. Lett., 71*, 229–240.

12. Cerling, T. E., (1992). Development of grasslands and savannas in East Africa during the Neogene. *Paleogeogr, Paleoclimatol, Paleoecol, 97*, 241–247.

13. Cerling, T. E., Wang, Y., & Quade, J., (1993). Global ecological change in the late Miocene: Expansion of C_4 ecosystems. *Nature, 361*, 344–345.

14. deMenocal, P. B., & Bloemendal, J., (1995). Plio-Pleistocene climatic variability in subtropical Africa and the paleoenvironment of hominid evolution: A combined data-model approach. In: Vrba, E. S., Denton, G. H., Partridge, T. C., & Burckle, L. H., (eds.), *Paleoclimate and Evolution with Emphasis on Human Origins* (pp. 262–288). New Haven, CT: Yale University Press.

15. deMenocal, P. B., (1995). Plio-Pleistocene African climate. *Science, 270*, 53–59.

16. Feibel, C. S., (1997). Debating the environmental factors in hominid evolution. *GSA Today, 7*, 1–7.

17. Feibel, C. S., Harris, J. M., & Brown, F. H., (1991). Neogene paleoenvironments of the Turkana Basin. In: Harris, J. M., (ed.), *Koobi Fora Research Project* (Vol. 3, pp. 321–370). Stratigraphy, artiodactyls and Paleoenvironments. Oxford: Clarendon Press.

18. Foley, R. A., (1994). Speciation, extinction, and climatic change in hominid evolution. *J. Hum. Evol., 26*, 275–289.

19. Gabunia, L., Vekua, A., Lordkipanidze, D., Swisher, C. C. III, Ferring, R., Justus, A., Nioradze, M., Tvsatchretidze, M., Anton, S. C., Bosinski, G., Jőris, O., de Lumley, M. A., Majsuradze, G., & Mouskhelishvili, A., (2000). Earliest Pleistocene hominid cranial remains from Dmanisi, Republic of Georgia: Taxonomy, geological setting, and age. *Science, 288*, 1019–1025.

20. Haile-Selassie, Y., Suwa, G., & White, T. D., (2004). Late Miocene teeth from Middle Awash, Ethiopia, and early hominid dental evolution. *Science, 303*, 1503–1505.

21. Hay, R. L., (1976). *Geology of the Olduvai Gorge, a Study of Sedimentation in a Semiarid Basin* (p. 203). Berkeley, CA: University of California Press.

22. Hays, J. D., Imbrie, J., & Shackleton, N. J., (1976). Variations in the earth's orbit: Pacemaker of the ice ages. *Science, 194*, 1121–1131.

23. Leakey, M. G., Feibel, C. S., McDougall, I., & Walker, A., (1995). New four-million-year-old hominid species from Kanapoi and Allia Bay, Kenya. *Nature, 376*, 565–571.
24. McBrearty, S., & Brooks, A. S., (2000). The revolution that was not: A new interpretation of the origin of modern human behavior. *J. Hum. Evol., 39*, 453–563.
25. McHenry, H. M., & Berger, L. R., (1998). Limb proportions in *Australopithecus africanus* and the origins of the genus Homo. *J. Hum. Evol., 35*, 1–22.
26. Partridge, T. C., Bond, G. C., Hartnady, C. J.H., deMenocal, P. B., Ruddiman, W. F., (1995). Climatic effects of Late Neogene tectonism and volcanism. In: Vrba, E. S., Denton, G. H., Partridge, T. C., & Burckle, L. H., (eds.), *Paleoclimate and Evolution with Emphasis on Human Origins* (pp. 8–23). New Haven, CT: Yale University Press.
27. Pickford, M., & Senut, B., (2001). The geological and faunal context of Late Miocene hominid remains from Lukeino, Kenya. C.R. Academy Science Paris, Sciences de la Terre et des planets. *Earth Planet. Sci., 332*, 145–152.
28. Poirier, F. E., & McKee, J. K., (1999). *Understanding Human Evolution* (4th edn., p. 386). Upper Saddle River, NJ: Prentice Hall.
29. Potts, R., (1996a). Evolution and climate variability. *Science, 273*, 922–923.
30. Potts, R., (1996b). *Humanity's Descent: The Consequences of Ecological Instability* (p. 325). New York: W. Morrow.
31. Ruddiman W. F., (2001). *Earth's Climate, Past and Future* (p. 465). New York: W. H. Freeman.
32. Semaw, S., Harris, J. W.K., Feibel, C. S., Renne, P., Bernor, R. L., Fessaha, N., & Mowbray, K., (1997). The oldest archaeological sites with an early Oldowan Industry from the Gona River deposits of Ethiopia. *Nature, 385*, 333–336.
33. Senut, B., Pickford, M., Gommery, D., Mein, P., Cheboi, K., & Coppens, Y., (2001). First hominid from the Miocene (Lukeino formation) Kenya. C. R. Academy Science Paris, Sciences de la Terre et des planètes. *Earth Planet. Sci., 332*, 137–144.
34. Shackleton, N. J., Berger, A., & Peltier, W. R., (1990). An alternative astronomical calibration of the Lower Pleistocene timescale based on ODP Site 677. *Trans. R. Soc. Edinb. Earth Sci., 81*, 251–261.

35. Stanley, S. M., (1995). Climatic forcing and the origin of the human genus. In: *Effects of Past Global Change in Life*. "National Research Council (U.S.)" Washington, DC: National Academy Press, pp. 233–243.

36. Tattersall, I., (1993). *The Human Odyssey: Four Million Years of Human Evolution* (p. 191). New York: Prentice Hall.

37. Vrba, E. S., Denton, G. H., Partridge, T. C., & Burckle, L. H., (1995). *Paleoclimate and Evolution, with Special Emphasis on Human Origins* (p. 547). New Haven, CT: Yale University Press.

INDEX

A

Alpine habitats 195
Alpine timberline 195
Ancient soils 176
Anemometers 2
Antarctic circumpolar current (ACC) 120
Anthropogenic records 183
Aquatic organic carbon estimation 156
Aquatic temperatures 175
Archaeological documents 204
Archaeology 81
Asteroids 63
Astronomy 60
Atmosphere-ocean exchanges 120
Atmospheric chemistry 134
Atmospheric composition 166
Atmospheric fluxes 114
Atmospheric general circulation model (AGCM) 113
Aurora borealis 50, 53
Australopithecines 206

B

Barometers 2
Benthic foraminiferal extinction (BFE) 133
Biblical flood 49
Biogeochemical procedures 73
Boosted monsoon flux 186
Boundary condition 114

C

Carbonate dissolution 132
Celestial mechanics 6
Classified stratigraphic methods 150
Climate 2
Climate changeability 169
Climate cycles 49, 51
Climate imposing mechanisms 156
Climate modeling study 151
Climate models 4, 9, 125
Climate variations 9, 22, 23
Climatic alteration 183
Climatic aridity 185

Climatic factors 168
Climatic memory effects 156
Climatic tolerances 96
Cognitive procedures 213
Comparative warming 152
Continental geometries 166
Coupled general circulation model (CGCM) 115
Cretaceous fossil angiosperms 97
Cyclic oscillations 189

D

De-glaciation 184
Dendroclimatology 168
Di-atom assessments 187
Diluvium 56
Distinctive traditionalism 46
Dynamical oceans 135

E

Earth processes 185
Earth systems 112
East Antarctic ice sheet (EAIS) 149
Ecological tolerances 97, 98, 99, 108
Enthalpy 101
Eolian movement 78
Erratic block 56
Euxinic waters 46
Evapotranspiration 3

F

Fossil accumulation 98
Fossil floral 99
Fossil mounds 185
Fossil organisms 98
Fossil plant associations 176
Fossil plants 96

G

GCM caliber models 116
General circulation models (GCMs) 150
Geochemistry 60
Geological model 67
Geological researches 204
Geologic statistics 204
Geomagnetic variations 51
Geothermal activities 166
GHG emissions 166, 169
Glacial theory 57
Glasshouse gas tests 14
Global climate models 204
Global warming 2, 42
Greenhouse gases (GHGs) 3
Gulf stream 112

H

Hanseatic league 75
Heat transfer 157
Hemispheric temperature circulation 157
Holocene epoch 182, 197, 198
Holocene fossil wood 195
Holocene timberline 193, 194, 199
Hominid fossils 210
Hominin-bearing sediments 204
Hominin fossil 205
Hot climate plants 212
Humanoid evolution 207
Hydrocarbons 74
Hydrogen isotopes 105

I

Ice cores 167
Ice-rafted detritus (IRD) 149
Infrared radiation 166
Inorganic marine sediments 121

Interactive ocean model 151
Isotope assessment 189
Isotope inconsistency 209
Isotopic composition 167, 169, 170, 171
Isotopic ratios 97
Isotopic spalling 62

L

Latitude 2
Leaf physiognomic approaches 100
Leaf physiognomy 100, 101, 102
Leaf venation density 100
Lunar cycles 52
Lunar nodal 52

M

Macrofossil analyses 191
Marine biota 157
Marine creatures 209
Marine documents 209
Meridional overturning circulation (MOC) 112
Meteorology 60
Microfossil 148
Milankovitch theory 173
Milky way 61
Millennial-scale variability 112
Moisture characteristics 104
Momentum fluxes 114

N

Neoproterozoic glaciations 136
Nodical cycle 76
Numerous species 205

O

Ocean basins 125, 126, 127
Ocean dynamics 116
Ocean gateway 119
Ocean general circulation model (OGCM) 115
Ocean heat transport (OHT) 112
Ocean modeling 113
Ocean paleomagnetic irregularities 67
Organic remnant 124
Organic systems 168
Oxygen isotope documentation 204

P

Paleobiology 3
Paleocene-eocene thermal maximum (PETM) 129
Paleocirculation direction 117
Paleoclimate documentation 210
Paleoclimate evidence 3
Paleoclimate information 213
Paleoclimate records 172
Paleoclimatic chronicles 6
Paleoclimatic signs 80
Paleomagnetic measurements 79
Paleomagnetic statistics 59
Paleomoisture parameters 102
Paleo-oceanographic modeling 113
Paleovegetation zone 99
Paleozoic periods 104
Periodic fortification 46
Plant assemblages 97
Plate-tectonic hypothesis 58
Plate tectonics 209
Pliocene experiment 154, 156
Pliocene scenario 155
Pliocene simulation 155, 156, 157
Pollen 190, 191, 194
Precessional milankovitch cycles 148
Primeval snowline 176

Primitive stone tools 208
Prognostic science 59

Q

Quasi-biennial oscillation (QBO) 76

R

Radiocarbon 182, 184, 192, 193, 194, 195, 197, 200
Radiocarbon dating 184
Radiometric dating 7
Rainy-season circulation system 187
Relict landscapes 185
River drainage systems 135

S

Sea surface temperatures (SSTs) 113, 153
Sedimentology 60, 81
Sessile organisms 96
Short-wave radiation 127
Social darwinism 205
Soil profiles 195
Solar radiation 52, 61, 62, 72, 76
Source water precipitation 173
Spatial climatic data 20
Stomatal density 100, 102, 107
Stomatal index 100, 102
Stratigraphy 60
Surface ocean computation 115

T

Taxonomic groups 97
Taxonomic identifications 97
Temperature variation 153
Temporal scales 166
Termeolization 79
Terrestrial climate 96
Terrestrial ice sheets 128
Thermal optimal level 185
Tidal currents 125, 126
Topography 2
Transantarctic mountain range 149
Treeless plant communities 190
Treeline communities 196
Treeline vegetation 192
Tree ring density 168

U

Uniformitarian 4
Universal scale energetics 158

V

Vein calcite 173
Vertical gradients 119
Volcanic activities 188
Volcanic emissions 166
Volcanic eruptions 77
Volcanism 133

W

Warm-dry cycles 183
Weather assessment 80
Weathering rates 149
Wind-blown dust 166
World-scale atmospheric temperature renovations 169